# DECODING THE LANGUAGE OF GENETICS

# DECODING THE LANGUAGE OF GENETICS

DAVID BOTSTEIN

COLD SPRING HARBOR LABORATORY PRESS
Cold Spring Harbor, New York • www.cshlpress.org

DECODING THE LANGUAGE OF GENETICS

Published by Cold Spring Harbor Laboratory Press, Cold Spring Harbor, New York
© 2015 by David Botstein
Printed in the United States of America

| | |
|---|---|
| Publisher | John Inglis |
| Acquisition Editor | Alexander Gann |
| Director of Editorial Services | Jan Argentine |
| Developmental Editor | Jane Alfred |
| Project Manager | Inez Sialiano |
| Permissions Coordinator | Carol Brown |
| Director of Publication Services | Linda Sussman |
| Production Editor | Kathleen Bubbeo |
| Production Manager | Denise Weiss |
| Cover Designer | Pete Jeffs |

*Library of Congress Cataloging-in-Publication Data*

Botstein, David.
    Decoding the language of genetics / David Botstein.
        pages cm
    Summary: "This is a book about the conceptual language of genetics. There is a need for special words and terms to deal with some of the essential abstractions in genetics; these are the focus of this book. It is intended to help readers with diverse interests and experience to think about genetic analysis in a more sophisticated and creative way"-- Provided by publisher.
    Includes bibliographical references and index.
    ISBN 978-1-62182-092-5 (hardback)
1.  Genetics. I. Title.
    QH431.B6355 2015
    572.8--dc23
                                    2015012574
10  9  8  7  6  5  4  3

All World Wide Web addresses are accurate to the best of our knowledge at the time of printing.

For a complete catalog of all Cold Spring Harbor Laboratory Press publications, visit our website at www.cshlpress.org.

*To the many students
who have taken my courses in genetics
at MIT, Stanford, and Princeton,
because it was they that made me think about
the importance of language in genetics in the first place*

# Contents

Preface, ix

Acknowledgments, xiii

1   The Basics, 1

2   Implicit Experiments and the Functional Gene, 17

3   Recombination and Linkage Mapping, 31

4   Pathway Analysis, 47

5   Regulation of Metabolic Pathways, 59

6   Phage and the Beginning of Molecular Genetics, 71

7   Transcription, Translation, and the Genetic Code, 87

8   Suppression Genetics, 105

9   Functional Suppression, 117

10   The Genetics of Complex Phenotypes, 131

11   Transcriptional Regulation of Gene Expression, 137

12   The Modular Architecture of Genes and Genomes, 153

13   Evolution Conserves Functional Sequences, 167

14  Human Population Genetics, 181

15  Inferring Human Gene Function from Disease Alleles, 195

16  What Is Next in Genetics and Genomics?, 207

Index, 211

# Preface

This is a book about the conceptual language of genetics. I was motivated to write it for the many people I have met who are interested in genetics, but who find the language confusing and intimidating. I think access to the powerful abstractions and beautiful ideas that underlie our understanding of how traits are inherited has become limited by the growth of well-intentioned, but unnecessarily complicated academic language and jargon. As a result, even relatively sophisticated biologists, chemists, physicists, and physicians have difficulty engaging with some of the ideas that underpin the power of genetic analysis. This is a problem for the future; more and more people will obtain the DNA sequences of their genomes in the next decade or two. These sequences cannot be understood without reference to the basic ideas of genetics.

I believe there is a need for special words and terms to deal with some of the essential abstractions in genetics; these are the focus of this book. At the same time, the field has also acquired, in the last century or so, many unnecessary and confusing words, sometimes as a result of misunderstanding, and sometimes as a result of academic discourse, where the drive for erudition tends to drive out straightforward exposition. In my 40 or so years of teaching genetics, I have come to believe that fundamental genetic concepts can be explained using simple language, without loss of rigor and with only a modest number of indispensable specialized words.

I organized this book around the specialized words that capture the most fundamental genetic ideas. My aim is to convey the meaning and utility of the essential and helpful ones as they are used today. In a few cases, significant disagreement remains among my colleagues over the use of a word or term: I try to make it clear when this is so. Because this is not a book about history, I highlight only those historical issues that remain the source of continuing confusion. The reader will find, nevertheless, that I have included some relatively detailed discussions of the origins and/or applications of some of the more fundamental concepts. I chose these not for the sake of history, but to aid with understanding by illustrating how such words and ideas work in practice.

In writing this book, I have assumed only minimal background knowledge of genetics. Readers should be able to understand and benefit from this book knowing only the most basic elements: the outline of Mendelian inheritance; chromosome mechanics (mitosis and meiosis) at the cartoon level; and, of course, the idea that functional genes are encoded as DNA arranged linearly along chromosomes. Many readers will have learned these basic concepts in a high school biology course. For those who, nevertheless, would like to refresh their recollection of these fundamentals, I highly recommend the first few chapters of *Genetics Notes* by the late, renowned human geneticist, James F. Crow.[1] In about 50 pages, Crow delivers all the basics a reader will need in clear and simple language. I provided a very few references to the original literature, restricting myself to those papers that are highly emphasized in the book and which are written in such a way that readers can appreciate them.

I hope this book will serve as a helpful companion for those thinking, reading, or writing about genetics; it is not intended to be a dictionary or textbook, let alone a history. No effort has been made to "cover" the field. I have made no attempt to transmit a set of biological facts—I tried to introduce just enough of them to illustrate important principles that underpin our understanding of genetics. In some cases, I have taken liberties. For example, scientists studying diverse organisms have developed different methods for naming their genes, resulting in a lot of species-specific genetic nomenclature. I have tried to avoid or to simplify naming conventions in order to spare the reader the confusion caused by these diverse ways of naming genes.

When I was a teenager, I spent time reading books about chess, bridge, or poker. These books were aimed at players with a wide range of experience. They assumed only knowledge of the rules and the basic moves. Like most readers, I could play these games, but by no means did I play them really well. I acquired and read these books in the hope of improving my game. My hope is that this book will serve, in an analogous fashion, to help readers with diverse interests and experience to think about genetic analysis in a more sophisticated and creative way.

To sum up, it is my hope that readers of this book will find it a useful guide to genetic ideas, regardless of their previous exposure to them. Genetics has come

---

[1]Crow JF. 1986. *Genetics notes: An introduction to genetics*, 8th ed. New York: Macmillan, New York.

to play a central role in modern understanding of biology. It is my dream that more of my professional colleagues will recognize that this means that we, as a discipline, should make our work more generally accessible by modernizing, clarifying, and simplifying the language we use and teach.

DAVID BOTSTEIN

# Acknowledgments

I am grateful to the many teachers, mentors, colleagues, and students from whom I learned what is important in genetics and genetics teaching. Although they are too numerous to mention individually, I owe them, and many of them will recognize their influence if they read this book. I am also grateful to my wife, Renee, and my family and friends for encouraging me to undertake writing this book, and to John Inglis and Alex Gann for accepting it for publication. The University of California, San Francisco, the Jackson Laboratory in Bar Harbor, and the Donnelly Centre at the University of Toronto each generously hosted me during parts of the sabbatical leave from Princeton that made this book possible. The staff of the Cold Spring Harbor Laboratory Press have been wonderfully efficient in turning the draft into a book. Finally, I am in debt to Jane Alfred for her excellent editing. She eliminated many errors and much awkwardness; any defects that remain are my sole responsibility.

CHAPTER 1

# The Basics

## Introduction

Genetics is the science of biological inference based on regularities in patterns of inheritance. It is a fundamentally analytical enterprise. It has become clear, especially in the last 50 years, that many important facts about biology can be uncovered using genetic analysis that cannot be learned in any other way. Genetic analysis is the best window we have into the workings of evolution, and few modern biologists would quarrel with the famous dictum of Theodosius Dobzhansky, a renowned 20th-century evolutionist: "Nothing in biology makes sense except in the light of evolution."[1] Genetic analysis depends on abstractions that have generated a specialized language with a private vocabulary. It is the essential concepts, language, and vocabulary of genetics that we explore in this book.

Patterns of inheritance can sometimes be observed in natural populations: For example, blue-eyed parents produce only blue-eyed children, whereas some brown-eyed parents can produce both blue-eyed and brown-eyed offspring. Some interesting inferences are possible from such observations in natural lineages, but they are quite limited compared to the inferences that can be drawn from quantitative analysis of the results of genetic experiments (see below) with well-chosen parents, brilliantly introduced in 1865 by Gregor Mendel using pea plants in his garden.[2]

Mendel's innovations were many, but the most important merit explicit mention:

---

[1] Dobzhansky T. 1964. Biology, molecular and organismic. *Am Zool* **4**: 443–452.

[2] Mendel G. 1866. Versuche über Pflanzenhybriden. *Verhandlungen des naturforschenden Vereines in Brünn, Bd. IV für das Jahr 1865*, Abhandlungen, 3–47. Mendel read this at the February 8 and March 8, 1865, meetings of the Brünn Natural History Society. William Bateson translated the paper into English in 1901. The best source for the paper today is MendelWeb (http://www.mendelweb.org), which gives both the original and Bateson's translation side by side.

---

1. The choice of garden peas. Like many plants, pea plants are normally self-fertilizing. However, gardeners can arrange that they are fertilized instead by plants of the same species that have different properties.

2. The choice of properties to study. Mendel selected properties, such as pea color and shape, that remain invariant for generation after generation when pea plants self-fertilize (a property called "breeding true").

3. The crossing of true-breeding plants. Mendel crossed plants that bred true for one trait with others that bred true for one or more different traits and then crossed the hybrid progeny with each other. This procedure resulted, at each generation, in progeny that differed from one another with respect to the inherited traits.

4. Counting carefully the numbers of the different types of progeny. Mendel was the first to recognize that the ratios of different types of properties to emerge from a cross might reflect the mechanism of inheritance. Accurate quantitation was thus of paramount importance. To this end, he produced and counted large numbers of pea progeny, the better to estimate the ratios of the different inherited properties.

Mendel's quantitative experiments marked the dawn of our understanding of genetic principles and opened up lines of biological inference that ultimately illuminated virtually every aspect of biology. Following Mendel's lead, today's experimental geneticists organize the study of inheritance patterns, and thereby the association of DNA sequences with traits in genetic model organisms, by choosing suitable parents and subjecting them to crossing and breeding schemes that will yield the desired genetic information most efficiently. The remarkable success over the last century of such genetic analysis in plants, flies, fungi, bacteria and their viruses, worms, and mammals all depended on the ability to choose the parents and follow the inheritance of their traits across several generations of progeny.

Human geneticists cannot simply choose parents or arrange crossings that will be informative. Instead, human geneticists adopted methods that allow one to search through the population for families that might yield information about particular inherited diseases. These methods rely on DNA technologies developed only recently, in the last two decades of the 20th century. Today, genetic inference is being increasingly used to discover and analyze inherited human disease-susceptibility genes by connecting them to differences in DNA sequence. We now can look forward to a day, in the relatively near future, when a routine medical examination will begin with the patient's genomic DNA sequence, which, as we shall see, is only interpretable through the lens of genetic analysis.

Whether one is analyzing arranged crosses in experimental animals or existing human families, an important feature of study design is an estimate of how much data will be required to achieve statistical significance, because the way in which DNA is inherited is fundamentally probabilistic. Mendel intuitively chose the number of progeny to count; today, sophisticated statistical estimates of power using computational methods are standard. It is worth noting that many of the theoretical advances in the fields of probability and statistics since Mendel's time were motivated by the continuing desire to make genetic inference more rigorously quantitative.

Extracting biological insight from crossing and breeding studies depends on our understanding and applying a small number of abstract ideas. Geneticists developed special words to represent these ideas, which, as we shall see, are not always readily conveyed in ordinary language. The invention of new words (or the appropriation of existing words) into a technical language that can express genetic abstractions began with Mendel himself and has continued ever since. In some ways, genetic concepts resemble those of physics and chemistry, although the diversity and range of biological individuality, as well as the role of chance, make the extraction of mathematical regularities from biological phenomena particularly challenging. In the end, mathematical formulae often turn out to be less useful in dealing with genetic abstractions than are well-defined words of special meaning, combined with a few stylized diagrams.

The abstractions and concepts that are the most important and useful to genetic analysis are independent of any organism. This is not to say that all organisms deal with their DNA in exactly the same way. On the contrary, they are very diverse, differing in such basic things as the number of copies of their genome their cells normally contain. The analytical ideas emphasized here apply to every organism, even though not every kind of experiment is possible in all of them. Specifically, some organisms (viruses, bacteria, plants, and some fungi) are easily manipulated in huge numbers, whereas others (whales, elephants, and, of course, humans) are not experimental organisms at all. Nevertheless, the basic analytical ideas of genetics apply to each.

In writing this book, I have placed a great deal of emphasis on "implicit experiments"—experiments that underlie the definitions of the most central abstractions of genetics. Each of these is based on real experiments that I have idealized for the purpose of exposition. Real experiments depend on details and realities, many of which are specific to the organism under study. I have deliberately left out such details, particularly those I have found to be unhelpful and even distracting for the general reader seeking to understand the general principles.

For similar reasons, I have tried to avoid illustrative examples that require substantial background knowledge in the biology of specific organisms. Readers will no doubt notice that I favor examples from human genetics. I did this for two reasons. First, most readers will know much more about human biology than they will about the biology of bacteria, flies, or nematode worms. Second, most of what we have learned in other organisms applies also to humans; often the basic genetics was first worked out in one or more experimental model organisms. The success of this approach to understanding human biology is, quite properly, the well-justified basis for the continued societal support of research with model organisms.

## Genetic Variation

All of genetics is based on studying variation in the genomic DNA sequences of individuals in a population. Individuals from every species naturally exhibit this variation; without it there could be no evolution. The level of genetic variation that exists in a natural population is a product of many factors. Notable among these are the rate of errors in DNA replication and the size and evolutionary history of the population. Some, but not all, of this DNA sequence variation has biological consequences and thereby contributes to the biological individuality of each member of a population. Of course, there are other equally profound contributions to individuality that arise from the environment and from the differences in the history of each individual.

Genetic inference is based on the outcomes of cross-breeding individuals of the same species, who differ both in their genomic DNA sequence and in one or more biological traits. Genetic analysis seeks to distinguish whether any differences in their DNA sequence influence one or more of these traits. Ultimately, for simply inherited traits (those that are attributable to the function of a single gene), this comes down to a conceptually simple question: Is a particular sequence variant inherited by progeny from one of its two parents always inherited with the particular trait that is also inherited from that parent? When a statistically secure connection is found between a sequence variant and a trait, we infer that this genetic difference causes, at least in part, that particular biological trait.

Experimental geneticists today no longer depend on natural variation for determining whether a DNA sequence is linked to a trait. This is because biologists have developed increasingly efficient methods of artificially inducing random alterations in DNA sequences for use in experimental organisms since the late 1920s. More recently, advances in DNA technology have made it possible for researchers to construct experimental organisms with specific DNA sequence changes at will. These advances have simplified the process of connecting DNA sequence differences with their biological consequences in these organisms. However, even these relatively straightforward experiments ultimately require the abstractions of genetic analysis for their interpretation.

## Structure and Function

What can one actually learn from studying patterns of inheritance? The great inferences began with Mendel in 1865. Although we recognize his historical importance, it is hard, after so many years, to appreciate the true nature of Mendel's contribution today, when faced with the huge body of knowledge about inheritance that has been discovered since his day. Mendel knew nothing about DNA, chromosomes, proteins, or even "information" in the modern sense (i.e., something that can be reduced to and faithfully transmitted as a string of binary digits). Nevertheless, it is possible to restate his central insight in modern language. Mendel realized that the experimentally reproducible patterns of inheritance of well-chosen traits in pea plants meant that each plant contains two copies of the DNA sequences that cause each of these traits, one inherited from each parent. He also realized that each plant passes one copy of that DNA information to each of its progeny. Each of the seven traits Mendel chose to describe in his paper behaved independently of the others in experiments involving more than one trait at a time.

Mendel called the causative DNA sequences "factors" (he used the German word *Anlage*, which has additional connotations in English, including "predisposition" and "potentiality"). There are two classes of questions one can ask about Mendel's factors that are of enduring interest. One concerns the *structure* of the factors. Today, we know that these factors are stretches of DNA sequence that reside at particular positions on chromosomes and that they can be tracked on the basis of their positions. The second concerns the *function* of the factors. Today, we know that the information encoded in these DNA stretches can affect the biology of an organism in many ways—most often by specifying a protein or RNA molecule that actually fulfills particular functions at prespecified times and places in the life of an organism.

The pea plant (*Pisum sativum*) that Mendel worked with is, in today's language, "diploid," meaning that the cells of the mature organism contain two copies of every chromosome (except for sex chromosomes) and thus two copies (not necessarily completely identical) of every genomic DNA sequence. As indicated above, the central inference of Mendel's paper is that during sexual reproduction, the reproductive cells of the plant (its pollen and ovules, also called the gametes) contain only one copy of every genomic DNA sequence; in modern terms they are "haploid gametes." This system of alternation between haploid and diploid is characteristic of most, if not quite all, eukaryotic organisms (including plants, fungi, humans, and animals). However, not all organisms are stably diploid; bacteria and viruses, notably, are normally haploid. Nevertheless, the basic concepts of genetic analysis apply to them all.

## Gene and Locus

Distinguishing between structural and functional issues is fundamental to thinking clearly about genetics. Some of the confusion between the two different aspects of Mendel's factors proved difficult to resolve. Thus, geneticists argued about the "nature of the gene" for decades. Today, most geneticists use "gene" and "locus" in ways that help to minimize this confusion.

*Gene:* In 1905, a Danish botanist and geneticist called Wilhelm Johannsen introduced the word "gene," defining it in a way that subsumed both the structural and functional aspects of Mendel's factors. Still today, the word "gene" is used quite generally in both a structural and a functional context. However, most of our thinking about genes today refers to function; the modern consensus about the functional interpretation of the gene dates back to the work of American geneticist and pioneering molecular biologist Seymour Benzer in the 1950s (whose work I discuss in more detail in Chapter 6). Thus, the use of the word "gene" to refer to sequence or positional information, which remains part of its meaning, has to be done with care. I will have much more to say about the functional definition of the gene in later chapters.

*Locus:* Many geneticists (including me) like to refer to a stretch of DNA as a "locus" when discussing its position in the genome, even when we know it's a named gene with a known function. The word "locus" today unambiguously refers to position, and not to function. It is often used to refer to groups of genes that reside in the same region of a chromosome but whose functions may or may not be related to each other.

## Genotype and Phenotype

The potential for confusion between functional and positional/structural considerations means it is important to keep separate in one's mind variation in DNA sequence from variation in biology among individuals. Sometimes, there is a causal relationship between the two; however, often there is no such causal relationship or it is a complex one. The earliest geneticists formalized this distinction (even though they knew nothing about DNA) by adopting two special words, "genotype" and "phenotype," that serve to keep separate DNA sequence variation and its biological consequences. In the following section, I explain the modern working definitions of these words.

*Genotype:* This word refers to the DNA sequence of an individual. Different individuals have different genotypes whenever they differ in their DNA sequence, whether or not this difference has any known biological consequence. Genotype includes, in principle, the entire genomic DNA sequence of an individual, but is often used in a way that refers only to a particular gene or locus or a limited subset of genes and loci. A quick way to remember this is that in the context of genetic analysis:

> ▸ *Genotypes are differences in DNA sequence that distinguish individuals from one another.*

*Phenotype:* This word refers to any and all differing biological properties of individuals, but not including differences in their genomic DNA sequences. Today, the word is used very broadly to include any traits of an individual. In genetic analysis, phenotype tends to be used more specifically to refer to traits when they might possibly be the biological consequences of specific differences in genotype between individuals. In this usage, phenotype, like genotype, is often used in a way that refers only to a particular trait or subset of traits. A quick way to remember this is that in the context of genetic analysis:

> ▸ *Phenotypes are the visible or measurable properties/traits that distinguish individuals from one another.*

The introduction of these words (by the same Wilhelm Johannsen who had invented the word "gene" a few years earlier) was important because, like Mendel, other early geneticists could only infer genotypes on the basis of the inheritance patterns of phenotypes. These words therefore made it easier to avoid confusion, and their use has persisted even with the emergence of our current understanding of genes as DNA sequences. It is largely because of these words that the distinction between genotype (what is encoded in the DNA) and phenotype (the visible or measurable potential consequences of a genotype) has remained the same since the days of Mendel, even though much has been learned (and not just about DNA) since.

Enough specialized vocabulary has now been introduced to allow a rigorous definition of what genetic analysis is all about:

▸ *Genetic analysis relates genotypes to their phenotypic consequences and vice versa.*

## Homozygote and Heterozygote

All crossing and breeding schemes involve the formation of cells that contain both of two alternative, parental DNA sequences of, for example, a locus or gene. In diploid organisms that reproduce sexually, every cell in the organism (except the sperm and the eggs, which are referred to as the "gametes") contains two entire genomes, one derived from each parent. In bacteria, which are normally haploid, cells sometimes contain two different copies of only part of their genome. Experimental geneticists, beginning with Mendel, have deliberately produced hybrid organisms, which differ at particular loci or at genes of interest. In 1902, immediately after the rediscovery of Mendel, an English geneticist named William Bateson introduced two special words, "homozygote" and "heterozygote," to distinguish between instances in which the two copies of DNA information in a cell are identical or different. These words are so useful that their usage quickly became standard and remains so today.

*Homozygote:* This word is used when the two alternative copies of genomic sequence present in a cell are identical. Geneticists often refer to homozygosity with respect to one or a few genes (or loci) and not necessarily with respect to the entire genome. Genetic variation is sufficiently common that the deliberate construction of an organism with two absolutely identical genomic sequences is challenging, even in model experimental organisms. Conversely, it is common, even in natural populations, for considerably long stretches of DNA (encompassing sometimes dozens of adjacent genes or loci) to be homozygous.

*Heterozygote:* This word is used when the two alternative copies of genomic DNA sequence are different. A cell or organism is said to be heterozygous whenever such differences exist. These differences can involve as little as one DNA base pair (bp) or as much as an entire chromosome. An important example of the latter is sex chromosomes. In humans, females have two complete copies of the X chromosome, but males have only one copy and contain a Y chromosome instead of the second X. Thus, only females can be homozygous at loci on the X chromosome. You should bear in mind that the two X chromosomes in females drawn from natural populations are generally not entirely identical because of the ubiquitous population-level variation in DNA sequence. Indeed, a substantial level of heterozygosity throughout the genome is the hallmark of natural populations, which are continually evolving.

*Hemizygote:* This is a more recent term that refers to a subset of heterozygotes—those entirely missing a particular stretch of genomic DNA, as opposed to having a second copy that contains a divergent sequence. Human males are properly referred to as being hemizygous for sequences on the X chromosome. The American geneticist Hermann J. Muller in 1935 extended the common usage of this word to instances in which a single gene or locus is completely deleted in one copy of a diploid genome, and this usage has become quite common. The word is not used in cases in which only a part of a gene is missing; in such cases, "heterozygote" is the correct word to use.

## DNA Variants: Mutations, Polymorphisms, and Alleles

DNA sequence differences have, over the years, acquired many diverse and sometimes confusing names. Before considering these, it is important to reemphasize that all DNA variants—whether they are natural or artificially induced, whether they affect only one DNA base pair or extend over many, whether they are common or rare—in the end are all just differences in DNA sequence.

*Mutation:* This word was introduced in 1901 by Hugo de Vries, a Dutch botanist and an early geneticist, to describe newly arising heritable changes in otherwise true-breeding varieties of plants. de Vries, one of the rediscoverers of Mendelism in 1900, recognized what he called mutant plants by noting new traits that turned out to be heritable in the ways described by Mendel. He inferred that there must have been a change in an underlying Mendelian factor. In modern language, we would describe this as inference of a change in genotype on the basis of an observed heritable change in phenotype. It is this change in genotype that de Vries called a mutation, and this has been the usage ever since.

Today, if we were to find such a mutation, and one that acts as a "dominant" mutation (see below), we would verify the above inference by sequencing the DNA of the mutant and comparing it to the DNA sequence of the (nonmutant) "wild-type" parent line. If, as we cross and breed this new mutant, we observe that whenever the new phenotype is found the variant sequence is also found, and vice versa, we have strong evidence that we have found the mutation that causes that phenotype. The breeding scheme is required because there is a considerable likelihood that sequencing the mutant would reveal the presence of more than one sequence change in the genome; where this is the case, it is the mutation that faithfully follows the phenotype in crosses that is the causative one. When modern oncologists sequence human tumors in search of causative mutations, this is exactly the problem they face, except that the crossing and breeding approach is not available to them.

In modern usage, the word "mutation" can refer to virtually any change in DNA sequence. Nevertheless, the word is most commonly used to refer to rare alterations in a DNA sequence that cause a phenotypic consequence in organisms, one that makes them distinguishably different from the organism as it occurs in the wild (hence the term "wild type"). Today, the word "mutation" by itself always refers to genotype; if one wants to refer to a phenotype associated with a mutation, the correct and unambiguous usage is "mutant phenotype."

For DNA sequence differences that derive from variation in natural populations, and especially those for which no phenotypic consequence has been found, I prefer the words "DNA polymorphism" or simply "DNA sequence variant." The use of these words offers some advantages. First, unlike "mutation," they are explicitly agnostic with respect to any potential phenotypic effect. Second, when they originate in natural populations, their characteristic phenotypes, if they have any, are actually all "wild type" (e.g., the different colors of flowers one might find in a field). In human genetics, we try to avoid referring to patients as "mutants," even when it is fully justified scientifically; the word carries unfortunate cultural connotations. For example, we now know that hemophilia A is caused in males by mutations on the X chromosome that affected individuals inherit from their mothers. However, the locus that determines the synthesis of the clotting protein that is affected or missing in hemophilia A is highly mutable, and about one-third of all hemophilic males contain a causative DNA change that is not found in the mother's DNA; these changes are new (rather than inherited) mutations. Nevertheless, we do not call hemophilic boys "mutants."

**Polymorphism:** This is a word with deep roots in biology. The *Oxford English Dictionary (OED)* found a use of this word by the British naturalist Charles Darwin in 1846, only 6 years after its first use in any publication, a reference to diversity among portraits of the queen (so not used in a truly biological context). The plain English definition in the *OED* is ". . .the occurrence of something in several different forms." This definition obviously applies straightforwardly to phenotypic variation in natural populations. Over the years, the word came to be used to describe variation in genotypes as well. Although the ambiguity between genotypic and phenotypic polymorphism is both unavoidable and acceptable up to a point, today it is usual to modify the word to indicate whether genotype or phenotype is meant. Of course, when the connection between genotype and phenotype is already well established, as, for example, in the case of blood group antigens in humans, then there is no problem: Both genotype and phenotype are polymorphic, and current usage reflects this.

*DNA Polymorphism:* This usage of "polymorphism" is unambiguous and refers specifically to differences in DNA sequence. As indicated above, it is useful as an alternative to "mutation" when referring to natural populations or to humans. Originally, some authors had required that a particular alternative sequence was present at a minimum frequency in a natural population for it to qualify as being a DNA polymorphism. I think these academic arguments are best left behind. Today, I use "DNA polymorphism" whenever I need to refer to specific DNA sequence differences in populations but do not want to use "mutation," "DNA sequence variant," or "DNA sequence change."

A method of mapping human disease loci using DNA technology, called linkage mapping (which I discuss further in Chapter 3), was first introduced in 1980. It takes advantage of the millions of DNA polymorphisms that are found at a relatively high frequency in human populations, most of which cause no phenotype. This abundance of functionally silent DNA polymorphisms also provides the basis for the various forensic methods that are used to identify individuals from samples of tissue or blood.

Several technologies that detect DNA polymorphisms came into general use during the 1980s as markers for genetic mapping. The DNA polymorphisms that each method detects were named differently, resulting in a zoo of acronyms. I do not think the differences between them are fundamental, but because these acronyms are used abundantly in the literature, I explain some of them here.

*RFLPs (Restriction Fragment Length Polymorphisms):* This approach detects sequence variants by using DNA-cutting bacterial proteins called restriction endonucleases, which recognize very short (4- to 10-bp) sequences of DNA, to cleave DNA into fragments of different lengths. These fragments can then be visualized as bands on gels that separate them by size—hence the name.

*VNTR or STR (Variable Number Tandem Repeat or Short Tandem Repeat) Polymorphisms:* This approach also uses restriction endonucleases to generate different-sized DNA fragments from polymorphic loci that feature tandem repetitions of short stretches of DNA: The variation is in the number of repeats. The RFLPs generated from such alleles can be particularly informative because these DNA loci often vary considerably between individuals, and many different lengths can be found in the population. As a result, markers for these polymorphic regions are used today in most forensic applications, notably in CODIS, the U.S. Federal Bureau of Investigation's Combined DNA Index System, which is used to analyze DNA profiles that can distinguish any individual human from all others except identical twins.

**AFLPs (Amplified Fragment Length Polymorphisms):** These polymorphisms are detected when cleaved fragments of DNA are amplified using polymerase chain reaction (PCR). This is a sensitive method that is also used to detect DNA polymorphisms for forensic purposes.

**SNPs (Single-Nucleotide Polymorphisms):** These polymorphisms consist of a single-nucleotide change and are detected using a variety of methods, most simply via the direct sequencing of DNA.

**Genotyping:** This word refers to the process of detecting DNA polymorphisms, regardless of the technique used. Essentially all DNA sequence differences can serve as genetic markers, and all classes of DNA polymorphisms can now be detected by direct DNA sequencing. This method is rapidly gaining favor given the precipitous reductions in the costs of new sequencing technologies.

**Allele:** This very useful word refers to mutants or variants in the DNA sequence at a single gene or locus. Like many of the important invented genetic words, its meaning has evolved over time as geneticists have wrestled to understand the nature of the gene. In 1902, William Bateson introduced the word "allelomorph" (shortened to "allele" in the 1920s) to describe the alternative heritable determinants (Mendel's factors) that segregate away from each other in heterozygotes to produce the Mendelian ratios observed in their offspring.

During much of the early history of genetics (i.e., until the 1950s), the relationship between "locus" and "gene" remained murky. Throughout this period, evidence for the existence of alternative DNA sequences derived from the same locus or gene (also called allelism) consisted of a confusing mixture of functional tests and genetic mapping. The classic literature must therefore be read with considerable care around this point. Much (but not all, unfortunately) of the confusion was dissipated by the work of Seymour Benzer, which will be discussed in more detail in Chapter 6. Earlier literature regularly assumes that all mutations or variants that map to the same locus control the same function and, conversely, that mutations and variants that control the same function will map to the same locus.

Today, geneticists use "allele" to indicate any of the many alternative DNA sequences that can occur at a single genetic locus (that is, a single limited stretch of DNA sequence in the genome). Because the definition of "allele" relates to DNA sequence, alleles are, therefore, features of genotype. However, experimental geneticists who work with model organisms often define "allelism" by using a functional test, called complementation (see Chapter 2), to show that putative alleles are alternative versions of the same functional gene. This inference assumes (with considerable justification) that all noncomplementing mutations that define a functional gene most likely reside at the same locus. In the discussion of complementation in Chapter 2, the reader will find that there are some rare exceptions to this generalization. Sometimes, direct DNA sequence evidence is used to associate differences in DNA sequence with a functional gene by showing that within these differences sit the known boundaries of a gene's sequence. This alternative form of evidence is particularly useful if the allele has no useful phenotype and thus cannot be tested functionally.

However, the DNA polymorphisms that can be used as genetic markers, especially in human genetics, generally have no phenotype and often do not sit within a functional gene. For this reason, allelism among such DNA markers is strictly a matter of differences in sequence (variants) along the same short stretch of DNA. Notably, some polymorphic DNA marker loci (the VNTR/STR subsets of RFLPs) are useful precisely because one finds multiple alleles at the same locus that are readily distinguishable from each other. For example, if such a locus contains a dozen repeats of a simple sequence, one can find up to a dozen distinguishable alleles for one locus, each containing a different number of repeats. This is what makes such polymorphic loci much more informative than SNPs in family studies or in forensic applications.

Functional genes, in any organism, can and frequently do display DNA polymorphisms in their sequences. Thus, several "wild-type" alleles may exist in nature, which will mostly have no known phenotype. Experimental geneticists use an informal convention whereby the unaltered, naturally occurring sequence of a gene or locus that is present in an organism used in genetic experiments is referred to as "wild type" to distinguish it from mutant alleles of the same gene. Readers might have already read in published genetic studies, for example, that a particular genetic mutation caused a phenotype that differed from the wild type, even though in nature the DNA sequence of the gene involved is polymorphic.

Much of experimental genetics concerns the manipulation and analysis of mutant alleles of functional genes. Sometimes, the mutant alleles are spontaneous or randomly induced mutations ascertained by their phenotypes. In more recent years, mutants have tended to be the result of the deliberate creation of DNA changes (often deletions) by genetic researchers. Mutant alleles of some genes can vary widely in phenotype, whereas others tend to have very similar phenotypes, differing only by small degrees. Rigorous genetic experiments require that any analysis of such similar phenotypes be carried out relative to the cognate wild-type allele. It is also good practice to compare a mutant pheno-type to a "null" allele (which today is typically created by carrying out a precise deletion of the entire gene).

Finally, much of human genetics focuses on mutations in human genes. As with mutations in the genes of experimental organisms, mutant alleles of some human genes can vary widely in phenotype. For example, more than half of all human tumors have mutations in the *TP53* gene. These are so-called "somatic mutations" that have arisen during the lifetime of the individual (as opposed to being inherited). Hundreds of different mutant *TP53* alleles have been found, ranging from simple changes in one base pair to deletions of the whole gene. These mutations produce diverse phenotypic effects, reflecting the diversity in the functions of the master regulatory protein that is encoded by the *TP53* gene. In addition, a few heritable DNA polymorphisms in the *TP53* gene are common in human populations for which no phenotype has been found. Other heritable *TP53* mutations cause a hereditary cancer predisposition disorder called Li–Fraumeni syndrome.

Accepting that one has to be sensitive about calling real people "mutants," in reality the scientific issues that concern genetic mutations in humans are identical to those in experimental organisms; therefore, using the same scientific language when discussing them is entirely reasonable.

For further information on human genetic mutations, I refer readers to several important databases on human mutations that are accessible through the Internet, where you will find abundant references to the human genetic mutations mentioned in this book, as well as to the relevant scientific literature.

### Databases on human mutations

- Human Gene Mutation Database (HGMD): http://www.hgmd.org
- Online Mendelian Inheritance in Man (OMIM): http://www.ncbi.nlm.nih.gov/omim/
- Genecards: http://www.genecards.org/
- Swiss Prot Diseases and Variants: http://swissvar.expasy.org

### INTRODUCTORY BIOGRAPHIES

**Theodosius Dobzhansky (1900–1975)** was an eminent geneticist and evolutionary biologist. In addition to a prolific research career in *Drosophila* genetics, he wrote several influential articles and books that brought together genetics and evolutionary biology, which at the time were regarded as separate disciplines.

**Gregor Mendel (1822–1884)** was the founder of the science of genetics. In 1865–1866, he published an account of his experiments with pea plants that convincingly demonstrated the basic laws of inheritance. His achievement was not recognized until 1900, when three geneticists (Carl Erich Correns [1864–1933], Erich Tschermak von Seysenegg [1871–1962], and Hugo de Vries [1848–1935]) reintroduced his work, each having rediscovered some of Mendel's findings.

**Wilhelm Johannsen (1857–1927)** was a Danish professor of botany. His interest in understanding the relationship between genetic and environmental causes of variation led him to coin the words "genotype" and "phenotype." He wrote a very influential textbook in which he first used the word "gene."

**William Bateson (1861–1926)** translated Mendel's foundational paper into English and became the leading advocate for Mendelism in England. He coined the word "genetics" and wrote the first textbook on the subject. He also introduced the word "epistasis" to describe the masking of the phenotype of a mutation in one gene by a second mutation in another. With Reginald Punnet, he discovered genetic linkage, although he did not accept the then nascent chromosome theory.

**Hermann J. Muller (1890–1967)** discovered the mutagenic effects of X rays in *Drosophila* using an elegant genetic technique to detect lethal mutations. As a student at Columbia University, Muller had participated in the early development of *Drosophila* genetics in the remarkable group led by Thomas Hunt Morgan. He was active at the interface of genetics and society. In his later years, he became a leading antinuclear activist warning of the threat of nuclear war and weapons testing.

**Hugo de Vries (1848–1935)** was a professor of botany in Amsterdam who observed the 3:1 Mendelian ratio in his own experiments in Amsterdam in the late 19th century, which fueled the rediscovery of Mendel's work in 1900. He gave the name "mutation" to suddenly appearing heritable variation.

**Charles Darwin (1809–1882)** was a British naturalist who introduced the most basic of biological principles: that species evolved from a common ancestor by "natural selection" of ever-fitter variants.

CHAPTER 2

# Implicit Experiments and the Functional Gene

Some of the most fundamental concepts in genetics are best defined by reference to what I call an "implicit experiment." The paradigm example is the idea of dominance and recessiveness, Mendel's singular contribution to the conceptual language of genetics and the heart of his contribution to the science.

Every use of the words "dominance" and "recessiveness" refers to the outcome of a specific implicit experiment involving organisms with specified genotypes. Sometimes the experiment is as simple as assessing a phenotype, but sometimes it can involve several crosses and other elements. In this chapter, I introduce a number of concepts that can be explained by using such implicit experiments. Like thought experiments favored by theoretical physicists since Einstein's time, thinking about implicit experiments in genetics can be illuminating even when there is no way to perform them in practice. A good example of this is the *cis/trans* test, used by Seymour Benzer to define the "cistron," which has become the basis of the modern definition of the functional gene (a test I explain in Chapter 6). Here, the implicit experiment is an idealized one that is so challenging to arrange in practice that only rarely has such an experiment actually been performed in its entirety.

## Dominance and Recessiveness

The use of an implicit experiment can make a rigorous definition of "dominant" and "recessive" straightforward. The experiment here is to compare the phenotype of a heterozygote ($\alpha/\beta$) at a given locus to that of homozygotes for each allele ($\alpha/\alpha$ and $\beta/\beta$). If the phenotype of the heterozygote resembles that of one of the homozygotes and not the other (let us say the $\alpha/\beta$ phenotype resembles that of $\beta/\beta$ and not that of $\alpha/\alpha$), the allele in common (in this case $\beta$) is said to be dominant. The other allele (in this case $\alpha$) is said to be recessive.

▶ *The phenotype of heterozygotes defines the dominance or recessiveness of alleles/genes.*

The same implicit experiment underlies every determination of dominance and recessiveness. It consists of obtaining heterozygotes and the homozygotes for a pair of alleles, ascertaining their genotypes, and assessing their phenotypes. From this perspective, one can see that Mendel carried out every element of this implicit experiment to show that round pea shape is dominant relative to wrinkled, and that yellow pea color is dominant relative to green.

1.  Mendel crossed pea varieties that had bred true for many generations to assure himself that each was, to begin with, homozygous. He expected that the hybrids of the cross would be heterozygotes, containing alleles from each of the parents.

2.  He was not satisfied with this expectation. He went further, and verified the genotypes by crossing progeny back to a parent wherever he could, noting particularly the reappearance of the recessive phenotype in further crosses. This was important, because now he knew for certain that these plants were heterozygous, because the recessive phenotypes reappeared in the progeny of each of them.

Mendel introduced the words "dominant" and "recessive" as a simple way of describing the phenomenon, which he found with each of the seven traits he studied. Recessiveness was a novel and interesting idea, and Mendel described it in this way: "The expression 'recessive' has been chosen because the characters thereby designated withdraw or entirely disappear in the hybrids, but nevertheless reappear unchanged in their progeny. . . ."

The method Mendel used to verify the genotypes of heterozygotes is often referred to as a "backcross." However, it quickly became clear to experimentalists that crossing putative heterozygotes to any strain homozygous for the recessive alleles in question (not necessarily a parent) suffices, especially when several loci are involved. Such crosses are called "test crosses."

The formal definition of dominance and recessiveness has remained the same from Mendel's time to the present day.

***Dominant:*** When the $\alpha/\beta$ heterozygote phenotype resembles the phenotype of $\alpha/\alpha$ and not $\beta/\beta$ homozygotes, the $\alpha$ allele is the dominant allele.

***Recessive:*** When the $\alpha/\beta$ heterozygote phenotype resembles the phenotype of $\alpha/\alpha$ and not $\beta/\beta$ homozygotes, the $\beta$ allele is the recessive allele.

Simple dominance and recessiveness are not the only possible outcomes of an experiment intended to demonstrate dominance relationships. Other common outcomes of the assessment of a heterozygote phenotype are discussed below.

*Codominant:* When the α/β heterozygote phenotype resembles both the α/α and β/β homozygote phenotypes, the α and β alleles are said to be codominant.

A classic example of codominance is the human blood type AB. Blood group phenotypes are assessed by the agglutination reaction of red blood cells to anti-A or anti-B serum. Red blood cells from people with the A blood type react to anti-A serum and not to anti-B serum; cells from people with the B blood type react to anti-B serum and not to anti-A serum. The genetic determinants of these two blood types are alleles of the ABO locus, named $I^A$ and $I^B$, respectively. Hetero-zygotes at the ABO locus who carry one copy of the $I^A$ allele, which determines the A phenotype, and one copy of the $I^B$ allele, which determines the B pheno-type, display both the A and B phenotype because their red blood cells aggluti-nate with either anti-A or anti-B serum.

It turns out that the $I^A$ and $I^B$ alleles are each dominant to a third allele $(i)$, which determines the O phenotype, which is characterized by red blood cells that fail to agglutinate when exposed to either anti-A or anti-B serum. The $i$ allele is thus recessive to either the $I^A$ or $I^B$ allele. The vast majority of the human pop-ulation falls into one of four phenotypic blood type classes (A, B, AB, and O) based on six possible genotypes:

Group A (who have the genotype $I^A/I^A$ or $I^A/i$)

Group B (who have the genotype $I^B/I^B$ or $I^B/i$)

Group AB (who have the genotype $I^A/I^B$)

Group O (who have the genotype $i/i$)

*Incomplete (or Partial) Dominant:* When the α/β heterozygote has a phenotype in between the phenotypes of the α/α and β/β heterozygotes, then both alleles are said to be incompletely dominant. The classic example of this is flower color in snapdragons: Heterozygote progeny of true-breeding red-flower strains and true-breeding white-flower strains have pink flowers.

The relationship between a functional gene and its biological function (or functions, as many genes have several) is often not simple, and dominance rela-tionships as a consequence can be surprisingly complex. The modern definition of dominance and recessiveness, based on the implicit experiment comparing the phenotypes of heterozygotes relative to their cognate homozygotes, offers a reliable means by which to navigate through this complexity.

The concepts of dominance and recessiveness defined in this way have two fea-tures that are crucial to the interpretation of dominance relationships. First, each assessment of dominance and recessiveness relates to a particular heterozygote, α/β in the example above. If we have multiple alleles of the same functional gene, their dominance relationships are only defined pairwise. If allele α is dominant to allele β, it does not necessarily follow that allele α will be dominant to other alleles of the same functional gene. Indeed, a variety of alleles have been found for most genes, and many have been studied intensively. Typically, many mutant alleles act recessively to the wild-type allele because they have lost their function, and only a few act dominantly to the wild-type allele. However, inferences based on such generalizations require experimental confirmation for each dominance relationship.

Second, the assessment of dominance and recessiveness is tied to a specific phenotype. Many genes are involved, directly or indirectly, in more than one biological function, and thus mutant alleles of the same gene can underlie several distinguishable phenotypes. The same mutation can have different dominance relationships with the wild-type allele, depending on which phenotype is being assessed.

▸ *"Dominance" and "recessiveness" are therefore terms that relate to phenotype and not to genotype. As such, we no longer speak of "recessive genes" or "dominant genes."*

The classic example of a single mutation with many phenotypes and many dominance relationships is the human β-globin gene. This gene (called *HBB*) specifies one of the protein subunits of hemoglobin, the carrier of oxygen and carbon dioxide in blood; it is one of the best-studied of all proteins. A single base change (A→T) changes the sixth amino acid in the β subunit of the hemoglobin subunit from the negatively charged and hydrophilic glutamic acid to the uncharged and hydrophobic valine. This mutation has many well-documented phenotypic consequences. Among them is the deadly human disease called sickle cell anemia, which is endemic in tropical and subtropical regions of the world. It is characterized by red blood cells that have an abnormal sickle shape (red blood cells are normally disk-like round cells); hemoglobin with altered electrophoretic properties; and resistance to malaria caused by the parasite *Plasmodium falciparum*. These are very diverse consequences of a single base change out of a total of the more than 3 billion bases that comprise the human genome!

What is notable and interesting about these diverse consequences is the difference in dominance relationships between the mutant allele (we will call it $HBB^S$) and the wild-type allele ($HBB^A$) for each of these phenotypes. Let us turn first to the recessive phenotypes.

**$HBB^S$ Recessive Phenotypes:** Sickle cell disease is a phenotype of $HBB^S$ homozygotes. $HBB^S/HBB^A$ heterozygotes, who are said to have the "sickle cell trait," are not anemic (a phenotype of homozygotes caused by altered red blood cells), and they have a normal life span. Few, if any, of a heterozygote's red blood cells sickle under normal conditions. Thus, these cells do not cause any of the pathology that is characteristic of the disease. However, extreme exertion and/or hypoxia can cause as many as 1%–2% of the red blood cells of heterozygotes to reversibly switch to a sickle shape. As such, heterozygotes occasionally experience quite severe symptoms of the disease and even death (e.g., in the military) as a result of sickling under the extreme stress of hypoxia. This should not, however, deter us from concluding that these are recessive phenotypes. It is not at all unusual for recessive traits to be a bit less than absolute. Recessive alleles quite often cause traces of their relevant phenotypes in heterozygotes, as they do in this case.

**$HBB^S$ Codominant Phenotype:** The electrophoretic mobility of the hemoglobin extracted from the blood of an $HBB^S$ homozygote (called hemoglobin S) differs from that found for the hemoglobin from an $HBB^A$ homozygote (called hemoglobin A). The hemoglobin from $HBB^S/HBB^A$ heterozygotes contains about half of each kind of hemoglobin. After electrophoresis (the separation of molecules by size or charge in an electric field), one finds two separate forms of hemoglobin protein, one at the electrophoretic position of the hemoglobin from $HBB^S$ homozygotes and another at the electrophoretic position of the hemoglobin from $HBB^A$ homozygotes. Thus, both of the electrophoretic phenotypes of the cognate homozygotes are seen in the heterozygote, which satisfies the definition of codominance. This difference in the charge of the protein is a result of the difference in the amino acid at position 6 of the β-globin amino-acid sequence: In hemoglobin S, uncharged valine replaces the negatively charged glutamic acid.

**$HBB^S$ Dominant Phenotype:** Sickle cell anemia is a serious disease; in the absence of medical treatment, homozygotes rarely survive to have children. What, then, could account for the high frequency of the $HBB^S$ mutation, and the fact that it has arisen several times independently in human populations? It turns out that the $HBB^S$ allele confers on the red blood cells of heterozygotes a marked resistance to the cause of malaria, growth of the parasite *P. falciparum* within them. This resistance is also observed in the red blood cells of $HBB^S$ homozygotes. Thus, the $HBB^S$ allele is dominant with respect to this phenotype.

This increased parasite resistance provides a marked (and well-documented) advantage in overall survival to $HBB^S/HBB^A$ heterozygotes over $HBB^A$ homozygotes in malaria-infested areas, accounting for the repeated selection for the $HBB^S$ allele despite the low survival prospects of $HBB^S$ homozygotes.

With respect to survival per se, the phenotype of $HBB^S$ homozygotes is sometimes referred to as a case of "overdominance," because heterozygote individuals actually have better survival prospects than do either $HBB^S$ or wild-type $HBB^A$

homozygotes. Like many geneticists, I am unenthusiastic about the use of words such as "overdominance" and "underdominance." Most of the time, it is better to parse a complicated phenotype into its simpler components, which is usually not so hard to do. In this case, survival of these two homozygotes is impaired for entirely different reasons. The dominant resistance to growth of the malarial parasite in red blood cells favors the survival of the heterozygote over the normal $HBB^A$ homozygotes, and the lack of sickling, anemia, and other disease phenotypes favors the survival of the heterozygote over the $HBB^S$ homozygote. In my view, there is no need for a fancy word here to describe these different phenotypic relationships.

What good is it to know whether a variant or mutant allele causes a dominant or recessive phenotype? Mendel invented the idea to account for the ratios he found in the progeny of intercrossed hybrids. The idea that the phenotype of some alleles might be masked in heterozygotes allowed him to explain these results very simply and neatly.

Our assessment of dominance and recessiveness remains important today. For example, to map a human gene to the genome (as discussed further below), we must take into account dominance or recessiveness because connections between a DNA sequence and a phenotype can only be made directly in homozygotes if the phenotype is recessive; it can be made in every carrier if it is dominant.

## Penetrance and Expressivity

The connection between genotype and phenotype for simple Mendelian traits can be as clear and direct in practice as it is in principle. This is especially true in experimental genetics, where every effort is made to control the environment and other variables, especially genetic factors other than the ones under study. Mendel went to considerable trouble to inbreed his plants extensively, so that he could be confident that the individuals he crossed differed as little as possible from each other genetically, except, of course, for the traits he wanted to study. Moreover, he used a single environment for his studies, his garden at the Abbey in Brno.

However, the clarity and simplicity of this connection—between genotype and phenotype—breaks down when we come to analyze natural populations. This is because these populations are outbred and therefore contain considerable background genetic variation. In studies of human genetics, because it is not possible to control this variation or the environment, all one can do is to try to account for them. Thus, our ability to document the connection between genotype and phenotype, even of simple Mendelian traits, becomes less than absolute when studying natural populations. The resulting uncertainty has come to be described by two different words: "penetrance" and "expressivity."

*Penetrance:* This word refers, specifically, to the fraction of individuals who have a particular genotype (like homozygosity for $HBB^S$) and who display a typical associated phenotype (like sickling). In this particular case, the penetrance is nearly 100%, but where other simple Mendelian traits are concerned, the penetrance can be much lower. For example, the penetrance of the *BRCA1* alleles, which predispose individuals to breast and ovarian cancer, averages ~50% at age 50 and ~85% at age 70. Knowing the penetrance of a phenotype is important, not just in human genetics but also more generally in quantitative and population genetics. Changes in penetrance almost always happen for a reason, typically the influence of the environment or of other genes that interact with the gene of interest. In the case of breast cancer and ovarian cancer, one might speculate that aging reduces the effectiveness of the mechanisms that overlap with, and that can compensate for, the cancer-limiting functions of the product of the *BRCA1* gene, resulting in a higher penetrance of the mutant alleles in older women.

*Expressivity:* This word is used to refer to qualitative variations in the phenotype of individuals with the same genotype at a particular locus. At first glance, this appears to be very similar to the concept of penetrance, and the two are often confused. However, most modern geneticists clearly distinguish them. Expressivity is used to refer to qualitative differences in phenotype. In contrast, penetrance of a mutation is used to refer to the probability that individuals of the same genotype will exhibit the same expected mutant phenotype.

A notorious example of variable expressivity is a human developmental disease called neurofibromatosis type 1. This disease is caused by dominant mutations in the *NF1* gene that manifest as a diverse range of phenotypes. Over all, the penetrance is high (most carriers of the gene have a recognizable phenotype), but the nature of the disease varies greatly. Phenotypes include relatively mild (but often disfiguring) skin lesions, cancers, musculoskeletal abnormalities (such as curvature of the spine), learning disabilities, vision disorders, and epilepsy. The likely reason for the highly variable expressivity is the involvement of the *NF1* gene product in the development of many different organs: The range of mutant phenotypes is believed to reflect the many biological contexts in which the *NF1* gene product functions.

I should emphasize that although variation in penetrance (or expressivity) of a particular genotype might exist, this does not in any way diminish the relevance and applicability of the concepts of Mendelian segregation, dominance, or recessiveness in understanding the inheritance of traits, even in humans. In experimental systems, variation in penetrance or expressivity is quite rare, because steps are taken to control both the background genetic variation and the environment.

When exceptional cases of variable penetrance and/or expressivity are found and studied in experimental systems, they are usually traced to unrecognized variability in these factors. The discovery of the phenomena of genetic suppression and of specific gene interactions were the result of such studies, as the reader will see in later chapters of this book.

There is no reason to think that human genetics is different; variation in penetrance and expressivity often occurs because we cannot control either the environment or the genetic background of the individuals we study, and the resultant variation has to be taken into account in quantitative modeling and genetic counseling alike.

For instance, the risks associated with inheritance of a mutation in the *BRCA1* gene present a very different problem for genetic counseling than the risks associated with the *NF1* gene. In the former situation, the uncertainties are about the likelihood of breast or ovarian cancer developing: The nature of the phenotype and the medical consequences are more or less the same—the counseled patient can know what to expect, but not when it will happen. With respect to neurofibromatosis type 1, the medical consequences can vary from the relatively mild to extreme disability, as well as an increased likelihood of developing cancer.

## Biological Interpretation of Dominance and Recessiveness

As is the case for many aspects of genetics, and for biology more generally, it is often possible to make useful generalizations that apply in most, if not quite all, cases. Here, I present the simplest biological interpretation of dominance and recessiveness of a mutant allele relative to its wild-type allele, one that is known to be true of many thousands of genes that have been studied in hundreds of different organisms. The reader should be aware, however, that not every case is simple, and sometimes the generalizations will not apply, for any of a number of reasons. There will be more discussion of these uncommon exceptions, and how to deal with them, later in this book.

*Mutations That Are Recessive to Wild Type:* A simple and intuitive model accounts for most mutations that are recessive to their wild-type alleles. The model posits that such mutations result in a loss of information in a sequence of DNA, which results in a loss of function. Most genes are stretches of DNA that encode a protein or other molecular product. In a wild-type organism, this product does something (it has a biological function, maybe several). The simple model says that the recessive mutation makes a change in the DNA code that impairs or eliminates this function, resulting in the mutant phenotype. This model is supported by experimental evidence; we know that most functional genes make proteins or RNA molecules that perform functions in the cell, and that mutations can, and generally do, eliminate or impair these products and thus their functions.

Now consider a heterozygote. One allele (the wild-type allele) encodes the fully functional product, whereas the altered code of a mutant allele encodes either a defective product or produces no product at all. Provided that this product's function can be fulfilled successfully with about half of what a cell would normally produce of it, a heterozygote (which produces half of the functional product) will resemble phenotypically the wild-type homozygote and not the mutant homozygote. In fact, we know that for the majority of proteins (especially enzymes, which act as catalysts) this condition is met.

In the defining implicit experiment in which we study the phenotype of the heterozygote, a loss-of-function mutation will cause a recessive phenotype. The phenotype of the wild-type allele, which retains the correct coding information, will be dominant to it. Thus:

▶ *The simple interpretation of recessiveness is a loss of function.*

This is true even when a gene and its product are multifunctional, resulting in several potential, distinguishable mutant phenotypes. As emphasized above, dominance and recessiveness must be assessed relative to each phenotype, separately. The presumption that loss of function is the basis for a recessive phenotype applies also to a multifunctional gene, but only relative to the phenotype that was assessed in the heterozygote.

***Mutations That Are Dominant to Wild Type:*** How do we account for a mutation that produces a phenotype dominant to that of its wild-type allele? The simple model for this requires that a mutation in a DNA sequence results in the encoded product acquiring new properties and/or function(s). Dominant alleles, in this model, overproduce or make unusually stable products, which results in a new phenotype. The heterozygote will display the phenotype caused by the altered or overproduced product and will thus resemble the mutant homozygote, whereas the wild-type homozygote will not have the new phenotype. Because it is much easier to lose information and function than it is to gain new function, it is not surprising that dominant mutations occur much less frequently than do recessive mutations. In the defining implicit experiment, a gain-of-function mutation will produce a dominant phenotype, and that of the wild-type allele will be recessive to it. Thus:

▶ *The simple interpretation of dominance is gain of function.*

***Null Alleles:*** It is illuminating to consider the special case of "null" alleles. The simplest null alleles are deletion mutations in which the DNA that makes up the functional gene is absent; today, such deletions can be constructed at will in most organisms. These alleles, therefore, cannot, by their nature, produce a function at loci where this function would normally be encoded in wild-type organisms. As one might expect, null alleles are always recessive, with one important exception (that of haploinsufficiency), which we consider next.

Null alleles (especially those caused by DNA deletions) are particularly important for understanding the different functions of a gene product in a cell, because all these functions will be absent in cells with such alleles. Over the years, collections (also called libraries) of such deletion alleles have been generated for a range of experimental organisms, and for every known functional gene. Today, the gold standard for identifying the true null phenotype of a gene is to use these constructed deletions. There are also other kinds of mutations that we can predict will cause loss of function or altered function, and I return to these in Chapter 7.

***Haploinsufficiency:*** Our simple model for recessiveness, as described above, is based on an important assumption: that about half of the normal amount of a gene product is sufficient to avoid a loss-of-function phenotype in heterozygotes. However, there are occasions when a phenotype occurs in a heterozygote that carries a wild-type allele and a known null allele that does not necessarily resemble the phenotype of the null allele homozygote. This is called a haploinsufficient phenotype. And its occurrence is not surprising considering that the heterozygote produces half as much of the required gene product as does the wild type, whereas the null-mutation homozygote produces none.

Haploinsufficiency, rare as it is in reality, has to be considered whenever a dominant mutation is encountered. For experimentalists, this possibility is easy to verify if one has null alleles of the same gene available to create a haploinsufficient mutant, which should have a phenotype that is dominant to wild type. If it does not, then the dominant mutation must have acquired a novel function.

***Dominant–Negative Alleles:*** This class of dominant mutations produces the recessive phenotype of a null allele in heterozygotes that also carry the wild-type allele. They often do so by encoding a defective product that interferes with the assembly of a multisubunit complex, one subunit of which is the product of the affected gene. Thus, the phenotype of heterozygotes in which one allele encodes a functional product and the other a product that interferes with function is because of loss of normal function, as occurs in mutant homozygotes. In reality, dominant-negative alleles are quite rare, but when they are found, they can be a very useful tool in experimental genetics.

In the 1930s, the American geneticist Hermann J. Muller introduced a number of special words, among them "hypermorph," "hypomorph," "amorph," "antimorph," and "neomorph," in an effort to classify and simplify discussions of allelic dominance relationships. In their time, they were brilliant and prescient, in that they foretold what molecular analysis would show decades later. However, Muller's words remain in use today in only a few areas of experimental genetics (notably developmental genetics). Like most modern geneticists, I feel that our understanding of the genetic concepts that underlie these terms is not made easier by the use of these words. This is particularly the case given that each term has been defined by the outcome of an implicit experiment, and in practice real experiments have revealed that amorphs are alleles with a complete loss of function; hypermorphs are alleles that overproduce a wild-type function (e.g., in which a gene has undergone duplication); hypomorphs are incompletely dominant mutations, which produce some function but less than that of a wild-type allele; neomorphs are dominant alleles that produce a new ("neo-") phenotype not found in the wild-type organism; and antimorphs are what we call today dominant-negative alleles. I avoid all of these terms, not least because molecular interpretations often exist that make a genotype/phenotype relationship clearer than can the use of Muller's specialized vocabulary.

## Complementation and the Definition of the Functional Gene

The modern definition of the functional gene rests on the idea of complementation, a logical extension of the implicit experiment that underlies dominance relationships. The history of the definition of a gene is complicated, not least because geneticists were uncertain and confused about the relationship between gene structure and function for decades. Fortunately, the modern use of complementation for defining the basis of what constitutes a functional gene can be understood quite simply, given two assumptions. First, complementation analysis assumes that each functional gene carries out a biological function by encoding some kind of product, the lack of which results in a phenotype. Second, complementation analysis is limited to mutations that act recessively relative to wild type, consistent with the idea that such mutants produce defective products or none at all. Dominant alleles of any kind cannot be used in complementation studies. If these assumptions are satisfied, a simple implicit experiment, closely related to the previously discussed test for dominance and recessiveness, can be used to define the functional gene.

***Complementation:*** Complementation is not just an abstract concept: It can be tested by constructing a compound heterozygote from two recessive mutations (let us call them *m1* and *m2*), each of which by itself produces a similar pheno-type when homozygous. Once it is constructed, we assess the compound hetero-zygote's phenotype, just as one does in dominance tests. The interpretation of this experiment, given the above assumptions, is intuitive: If the *m1*/*m2* heterozygote shows the same phenotype as the homozygotes of each of the mutations (i.e., *m1*/*m1* and *m2*/*m2*), then there remains an absence of function. Therefore, *m1* and *m2* must each be defective in specifying the *same* functional product. In this event, the mutations are said to *fail* to complement. If the heterozygote shows the wild-type phenotype, then function has been restored in the heterozy-gote, meaning that *m1* and *m2* must each be defective in specifying a *different* functional product, and the mutations are said to complement.

A definition of the functional gene arises from this implicit experiment: It is the stretch of DNA information that specifies the functional product (usually a protein, but sometimes an RNA). Noncomplementing mutations that act reces-sively to wild type lie in the *same* functional gene; complementing recessive mutations lie in *different* functional genes. Standard genetic notation reflects this conclusion: once *m1* and *m2* are known to complement, the heterozygote is no longer written as *m1*/*m2* (which implies allelism, i.e., that they are alleles of the same gene) but instead is written to reflect that *m1* and *m2* are alleles of different genes, as follows:

$$\frac{m1}{+} \frac{+}{m2},$$

where "+" indicates a wild-type allele. This notation preserves the parental con-tributions of the heterozygote: The *m1*/*m1* homozygous mutant parent contrib-uted a wild-type allele of *m2* and the *m2*/*m2* homozygous parent contributed a wild-type allele of *m1*.

When mutations or alleles fail to complement each other, they are often said to fall into the same "complementation group." In earlier times, when allelism was determined solely by an allele's location in the genome, functional alleles were sometimes referred to as "pseudoalleles." I use the words "gene" and "allele" freely in their functional sense, using the more explicit "functional gene" and "functional allele" when there is a risk of confusion.

Occasionally, two alleles will also partially complement each other, as I explain in my discussion of compound heterozygotes below.

***Compound Heterozygotes:*** Compound heterozygotes are organisms that carry two different mutant alleles of the same functional gene. If both mutations are null mutations, then in functional terms a compound heterozygote will resemble a homozygote for either of the alleles. Sometimes, however, some function is restored by one or both mutant alleles in a compound heterozygote, in which case the two alleles are said to display "intragenic" complementation. This situation is sometimes referred to as "interallelic" complementation, which is so confusing to many that I do not recommend this usage.

*Exceptions to These Rules:* It should be noted that there are some important exceptions to the simple rules I have outlined here. Sometimes, recessive mutations in the same gene result in different phenotypes, and occasionally, recessive mutations in the same gene can complement each other—that is, they show "intragenic" complementation. Our current ability to construct true null alleles (which fail to provide any function) using modern genetic techniques makes it much easier to detect and explain these exceptional cases. For example, we can compare the phenotypic effects of a constructed deletion allele, which is missing all of the functional gene and the information it encodes, with individual recessive mutations, which might be missing only some of a gene. Fortunately, it turns out that most genes and most recessive alleles are well behaved, so the simple definition of functional genes based on complementation generally works pretty well.

To summarize:

- Complementation sorts recessive mutations into functional genes.
- A functional gene refers to the stretch of DNA within which recessive alleles generally fail to complement each other.
- A functional allele refers to recessive mutations that fail to complement each other.

# CHAPTER 3

# Recombination and Linkage Mapping

The definition of the functional gene in the previous chapters provides a modern view of the functional features of Mendel's factors. This chapter concerns their structural features. One might imagine that with modern DNA technology all that is required is to collect many recessive mutants that share a similar phenotype, find a group whose members all fail to complement each other pairwise, and sequence their individual genomes. The structure of genes turns out to be much more complicated than just stretches of contiguous sequence that encode functional molecules. Even as sequencing becomes cheaper and easier, the interpretation of sequence remains a problem.

A more fundamental problem with the "just sequence everything" approach is that the resulting connection between the region in which mutations lie and their associated phenotypes would be merely statistical, and really convincing if, and only if, large numbers of mutant genomes were successfully sequenced. Even then, as we will see, such an approach will not compete in simplicity and reliability with the more robust classical approaches.

The most important classical methods involve the way in which genes are passed on from one generation to the next. Each offspring inherits one chromosome from each parent. However, during the process by which gametes (eggs and sperm) are generated, the parental chromosomes undergo a process called "recombination," or "crossing over," which results in the gametes receiving chromosomes that are partly derived from each of the chromosomes the parents themselves had inherited. The classical methods for tracing genes through the generations are based on the observation that sequences that lie near each other on a chromosome are "linked" and are much more likely to be inherited together than are sequences that are distant from each other or on different chromosomes altogether.

Mendel introduced the idea of random assortment of his "factors" during reproduction; this was his explanation for the ratios he so carefully measured.

This process only later became associated with the idea that the determinants of different traits are carried on different chromosomes. It was researchers in the famous "fly room" at Columbia University, a genetics laboratory led by the pioneering American geneticist, embryologist, and evolutionary biologist Thomas Hunt Morgan, who firmly established the idea that genes lie on chromosomes. They also introduced linkage and recombination as a way of ordering genes based on their chromosomal locations. Specifically, it was American geneticist Alfred Sturtevant, then a second-year college student working in the fly room, who showed in 1913 that quantitative analysis of the frequency of recombination between genes on the fly X chromosome could be used to map them relative to one another.[1]

## Random Assortment, Linkage, and Recombination

The seven traits studied by Mendel all shared the same property: They assorted independently of each other when hybrid plants heterozygous for two traits (dihybrids) were crossed. Because one of the alleles for each trait was dominant, Mendel observed the famous 9:3:3:1 ratio in every combination of traits he tested (as explained below). How could these phenotypic ratios be explained? In Mendel's analysis of dihybrid crosses (AaBb × AaBb, where alleles A and B are dominant and alleles a and b are recessive), he inferred, using modern language, that all four possible gametes (AB, Ab, aB, and ab) were produced with equal frequency and that fertilization was random and independent of genotype. The question of how many genotypes of each kind will be represented in the hybrid's progeny is now reduced to the probabilities of the various combinations that can be produced in the gametes. Of the 16 genotypes that could arise, there is only one way to produce a plant displaying both recessive phenotypes (aabb); three ways to produce each kind of plant with only one of the recessive phenotypes (AAbb, Aabb, aAbb, aaBB, aaBb, and aabB); the remaining nine are all combinations that produce plants with both of the dominant phenotypes (Fig. 3.1). A cross that involved three independently assorting markers, analyzed in the same way, confirmed that all the possible gametes were produced with equal probability.

All of this, of course, preceded the discovery of meiosis and our understanding of the role of chromosomes (or DNA) in heredity. In the earliest days of research on chromosomes, the temptation was to imagine that each of the traits that assorted independently might reside on a different chromosome, because it

---

[1] Sturtevant AH. 1913. The linear arrangement of six sex-linked factors in *Drosophila*, as shown by their mode of association. *J Exp Zool* **14:** 43–59.

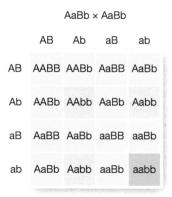

**Figure 3.1.** A diagram called a Punnett square that illustrates all the possible combinations of gametes in a plant dihybrid cross. Of the 16 genotypes that can arise from a cross between the genotypes *AaBb* × *AaBb*, there is only one way to produce a plant with both recessive phenotypes (pink) and three ways to produce a plant with one of the two recessive phenotypes (purple, green). The remaining nine are all combinations that produce plants with both of the dominant phenotypes (blue). See also Figure 14.1.

appeared from cytological studies that chromosomes also assorted independently. Mendel's peas (*Pisum sativum*) have seven chromosomes, and Mendel studied seven traits. Even today, if one looks in elementary genetics books (or searches on the Internet), one finds this simple but erroneous explanation. Three of the loci Mendel studied actually reside on chromosome 4 of *Pisum*, two reside on chromosome 1, and one each resides on chromosomes 5 and 7. It turns out that Mendel did not carry out all possible crosses. Had he done them all, he might possibly have found the linkage between the two traits whose loci reside on chromosome 4, the only two loci that turn out to be linked closely enough for him to have detected such physical linkage.

In any event, soon after the rediscovery of Mendel's work in 1900, the number of independently assorting traits exceeded the number of chromosomes in most organisms under study. This meant that there must be alternate mechanisms for reassorting loci during gametogenesis that worked even when the loci reside on the same chromosome.

*Recombination:* The earliest geneticists of the 20th century introduced this word into the genetics lexicon. The first published use cited by the *Oxford English Dictionary* was in 1903. At that time, it was very likely used to cover all the ways in which the specific alleles of two different genes that were introduced from one parent could be redistributed in new combinations in their gametes.

In cases in which these alleles reside on different chromosomes, their redistribution would result in all possible allele combinations being present in the gametes, just as Mendel had found. However, when alleles of two different genes reside on the same chromosome, there must be another explanation for the appearance of recombinant progeny. Somewhere in the process of producing gametes, there must be an opportunity for chromosomes to exchange information.

*Meiosis:* This is the name for the cell biological process that allows diploid eukaryotic organisms to produce haploid gametes. Meiosis is sometime referred to as "reduction division" for this reason. As shown in Figure 3.2, meiosis begins with a cell in which the chromosomes replicate (making a total of four copies of each) and the homologous chromosomes pair and have an opportunity to exchange information by a process called "crossing over." Thereafter, the cell divides twice (with no further replication), resulting in four haploid gametes. Crossing over can be visualized cytologically in some organisms, and it indeed was observed well before its genetic significance became clear.

So, during meiosis, segments of DNA can be swapped between paired chromosomes, breaking up the pattern of alleles from that found in either parent. If crossing over is very frequent, all possible combinations of parental alleles will appear in the gametes with equal frequency, just as they would when loci reside on different chromosomes and are assorted independently. If, however, crossing over occurs only occasionally, then the parental pattern of alleles is often not

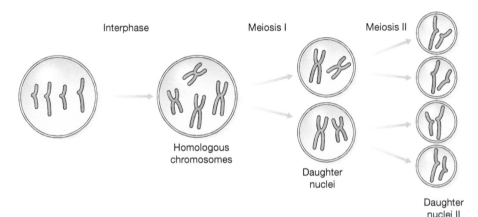

**Figure 3.2.** *Crossing over during meiosis. During the first meiotic division, called meiosis I, homologous chromosomes (shown in red and blue) pair and can exchange information through a process called "crossing over." During meiosis II, the paired chromosomes separate and the cells divide a second time.*

broken up by recombination and so will be more likely to appear in the gametes of their progeny.

Although some textbooks restrict the word "recombination" to mean the reassortment of alleles of genes on the same chromosome or DNA molecule, the word is actually more useful when it covers all mechanisms of reassortment; like most modern geneticists, I use it in this latter way.

**Linkage:** Two loci are said to be linked when the parental arrangement of the alleles at the two loci is favored over the recombined arrangement in the gametes. The degree of linkage (measured as the frequency of recombination, i.e., the frequency with which crossovers disrupt linkage) provided the first means by which to measure the physical distance between two loci on the same chromosome. The first genetic maps were made with fruit flies (*Drosophila melanogaster*) by Sturtevant and others in Morgan's group (see introductory paragraphs for more information on these early fly geneticists). Thereafter, this method for ordering genes on chromosomes was quickly applied to other experimental organisms.

## Measuring Distance between Loci by Recombination

The frequency of recombination between two loci is a quantitative parameter that provides a general basis for mapping loci. The experiment that yields recombination frequency between two loci superficially resembles, but actually fundamentally differs from, the tests of dominance and complementation.

Like the complementation test, a measurement of recombination begins with a heterozygote that differs at two loci. However, instead of assessing a phenotype, the heterozygote is allowed to make gametes. We then assess how many of its gametes are "recombinant gametes" (those that contain allele combinations not found in the parents) and how many are not (those that contain combinations of alleles found in the parents). No assessment of phenotype of this double heterozygote is involved. This is not a test of function, but rather a measurement that sits squarely in the realm of gene/genome structure. Before the advent of recombinant DNA technology and DNA sequencing in the late 20th century, recombination mapping was the only general method available for assessing the relative positions of loci on genomes.

In what follows, I largely restrict the discussion to crosses involving just two loci (so-called two-factor crosses), although the reader should be aware that crosses involving many loci are possible and sometimes yield considerably more information. The advantages and complications of multifactor crosses, nevertheless, will remain outside the scope of this book.

*Two-Factor Cross:* A two-factor cross is one that begins with the crossing of two diploid parents that are homozygous for distinguishable alleles of two loci (*A* and *B*) and that produces a double heterozygote. To simplify, let us use an example in which all the alleles are codominant, so we can assess their presence directly in heterozygotes. Using numbers to indicate alleles, the cross might be [*A1/A1 B1/ B1*] × [*A2/A2 B2/B2*].

The double heterozygote is conventionally written

$$\frac{A1}{A2} \quad \frac{B1}{B2},$$

with the alleles derived from one parent above the line and the alleles derived from the other parent below the line, a nomenclature convention that serves to preserve the parental arrangement of the alleles.

The four possible haploid gametes that can be produced are *A1 B1*, *A2 B2*, *A1 B2*, and *A2 B1*. Two of these gametes retain the parental arrangement (*A1 B1* and *A2 B2*) and two do not (*A1 B2* and *A2 B1*). The latter are referred to as "recombinant gametes" and the former as "parental gametes." As indicated above, when there is a statistically significant excess of parental gametes, one concludes that the two loci *A* and *B* are linked.

The simplest measure of recombination frequency is defined as follows:

$$F_{\text{rec}} = (\text{number of recombinant gametes})/(\text{total gametes}).$$

This parameter can vary from 0 to 0.5. When there is no recombination (i.e., when there is complete linkage), $F_{\text{rec}}$ is 0, and when there is abundant recombination (i.e., no linkage), $F_{\text{rec}}$ is 0.5 (meaning that parental and recombinant gametes are equally frequent). This is the same result that one expects when the genes reside on different chromosomes. So the original definition of recombination serves us well, making random assortment the limit of a continuum of recombination frequencies, irrespective of the mechanism involved.

The frequency of recombination is directly related to the physical distance between the recombining loci along the DNA molecule or chromosome on which they reside. A simple model that accounts for this general observation is that the recombination frequency reflects the probability of a crossover, and the more DNA there is between loci, the higher this probability will be. In most organisms, recombination is frequent enough so that loci at opposite ends of a chromosome arm are usually unlinked (i.e., $F_{\text{rec}}$ is 0.5, meaning that recombination is so frequent that the number of recombinant gametes equals the number of parental ones, and no remnant of linkage remains).

▶ *The frequency of recombination allows mapping of loci relative to each other, without reference to their function.*

The metric actually used for recombination (usually denoted $\theta$) is not simply $F_{rec}$, for a variety of reasons, the most important of which is that $F_{rec}$ values are only additive over very short distances because they do not account for the possibility of double crossovers. In 1919, the British geneticist and mathematician J.B.S. Haldane introduced a mapping function that takes double crossovers into account in a simple way, assuming that the frequency of double crossovers is the product of the frequency of the individual single crossover, as if they were completely independent events. Haldane's mapping function allows the calculation of a distance based on $F_{rec}$ that is additive and also, as it turns out, reasonably proportional to the physical distances (i.e., number of DNA base pairs) that separate the genetic loci. More-precise mapping functions have been devised since for different organisms, because each turns out to have somewhat different constraints on the occurrence of double recombination events that make them not quite independent of each other; for instance, in some situations, the presence of one crossover increases or decreases the frequency of a second crossover.

*Centimorgan:* The centimorgan (cM) is the unit of genetic distance that can be inferred from recombinant frequency. It provides a means of measuring the genetic distance that lies between loci on the same chromosome. It was named in honor of Thomas Hunt Morgan. At low recombination frequencies, $F_{rec}$ and the distances ($\theta$) calculated using Haldane's (or any other) mapping function are the same, with 10 cM corresponding to 10% recombination (i.e., $F_{rec} = 0.1$).

The physical amount of DNA represented by 1 cM varies from organism to organism, because organisms vary greatly in chromosomal structure and DNA content. On average, in most eukaryotic organisms, one crossover occurs per chromosome arm. In some organisms (such as in humans), a chromosome arm contains many millions of base pairs, whereas in others (such as in yeast), a chromosome arm might consist of as few as 100,000 bp. Thus, organisms with small genomes have many more crossovers per unit of DNA, resulting in smaller lengths of DNA per cM. For example, 1 cM in yeast corresponds to about 1000 bp (a kilobase, kb), whereas 1 cM in the human genome corresponds to about 1 million bp (a megabase, Mb). This more or less reflects the difference in total nuclear DNA between the two organisms: (12 million bp, or 12 Mb, in yeast, and 3 billion bp, or 3000 Mb, in humans).

## Defining Loci by Failure to Recombine

Given that a functional gene or locus is a stretch of DNA that is located some-where in the genome, one can use recombination as a means to assess whether two mutations might represent the same gene or locus. This is because alleles at the same locus should recombine very rarely or not at all, being, by hypothesis, located in the same stretch of DNA. According to our simple model, the probabil-ity of a crossover when there is little or no DNA between loci will be very low or even zero. Conversely, alleles at different loci should recombine very frequently when they are located at a distance from each other, and less frequently when they are closer together. In the early days of genetics, before the modern defini-tion of the functional gene, measuring the frequency of recombination formed the experimental basis for defining allelism. For simple base-pair changes, some recombination between alleles of the same functional gene should be detectable, as I discuss in Chapter 6, but in general it occurs much less frequently than does recombination between alleles of all but immediately adjacent loci.

## Linkage Mapping in Human Families

It should now be clear that correlation provides the mathematical basis for recom-bination mapping: Linked loci will be inherited in a correlated fashion, with the strength of the correlation related inversely to the physical distance between loci on the DNA. $F_{rec}$ is actually a kind of anti–correlation coefficient. It should also now be clear that our implicit recombination-mapping experiment happens whenever any animal (including humans) reproduces. Human gametes are, like those of yeast or flies, the products of meiosis, during which recombination occurs. Thus, any natural population is the result of a continuing series of recom-bination tests.

DNA polymorphisms provide a powerful means by which the inheritance of human DNA segments can be followed.[2] Each of us is heterozygous at about $1/1000$ of our 3.3 billion bp; some of these polymorphisms are common in every population, and others are rare. All of them, however, can be identified from tiny samples of DNA extracted from our cells, forming the basis for the forensic labo-ratory techniques that have become the staple of crime shows on television. Today, a standard source of genomic DNA for genetic testing purposes is saliva. Collecting a saliva sample is a simple and noninvasive procedure suitable for use

---

[2] Botstein D, White RL, Skolnick M, Davis RW. 1980. Construction of a genetic linkage map in man using restriction fragment length polymorphisms. *Am J Hum Genet* **32:** 314–331.

in humans. Of course, DNA can also be obtained from blood or from tissue samples. It is through the use of these techniques, and others, that we can study the inheritance of genetic loci.

The first step in studying inheritance is to identify genetic loci that differ between parents. For genetic purposes, the most abundant, convenient, and powerful differences to find are DNA polymorphisms. If such DNA polymorphisms differ from each other in each parent, then their inheritance can be tracked in the offspring; if not, then this particular polymorphism will not yield information about its inheritance. The different polymorphisms that can be found at a locus are called marker alleles. In the accompanying illustration (Fig. 3.3) of a genetic family tree, I have made up genotypes for a highly polymorphic locus $A$, such that both parents are heterozygotes for marker alleles of this locus and all four alleles can be distinguished from one another. All the possible genotypes of their offspring are shown below; the sex of the progeny was assigned at random—I will not consider the special case of genes on the sex chromosomes.

DNA polymorphisms that produce many alleles exist, as exemplified by RFLPs or VNTR polymorphisms (see Chapter 1). During the first decade of human gene mapping, researchers preferred using these types of markers to simple DNA polymorphisms because, unlike SNPs (see Chapter 1), they often have readily discernible multiple alleles. Today, we have the DNA sequence of the human genome, from which we can identify multiple alleles using an ensemble of SNPs from the same immediate vicinity. With so many and such highly polymorphic loci to hand, we can verify the Mendelian inheritance of loci in humans using just one or a few families, as I explain below.

**Mendelian Segregation:** Mendelian segregation is sometimes referred to as Mendel's First Law. By "segregation," geneticists mean the inheritance by offspring of

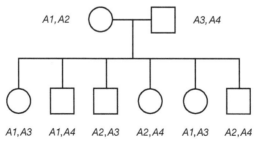

**Figure 3.3.** The segregation of a polymorphic locus, A. A schematic of a hypothetical family in which both parents are fully informative at a highly polymorphic locus, A. The segregation of the polymorphic alleles of A from each parent is illustrated in their progeny. In this type of family tree, males are represented by squares and females by circles.

one allele from each of the two alleles present at every locus in each diploid parent. In the example shown in Figure 3.3, each child inherited either *A1* or *A2* (not both) from the mother and either *A3* or *A4* from the father. Thus, the alleles in each parent were "segregated" away from each other. This is a consequence of meiosis, the process that produces gametes (see Fig. 3.1).

*Informativeness:* An informative cross is one that provides information about inheritance in a given situation. For example, using the family scenario described in Figure 3.3, if neither parent had been heterozygous at locus *A*, then neither they nor the cross would be informative. Furthermore, for a cross to be informative, it is also important for the two parents to carry alleles that do not introduce unresolvable ambiguities. For example, if both parents had the genotype *A1,A2*, then it would be possible to decide that a child with genotype *A1,A1* must have inherited the *A1* from both parents; similarly, an *A2,A2* genotype would mean that this allele had been inherited from both parents. However, in the absence of other information, it is not possible to decide which parent provided which allele to a child with the *A1,A2* genotype; such a child is not informative.

Now consider another, very similar family in which only one parent (in this case the mother) is informative (Fig. 3.4). I have made the father homozygous at all the loci we are going to discuss. I have done this only for the purpose of exposition; in real life, one might have to deal with the genotype of both parents, especially if they have some alleles in common, as mentioned above. In this family, I have included the genotype of another polymorphic locus *B*, at which the mother is heterozygous for two marker alleles (*B1* and *B2*) and the father is homozygous for a third marker allele (*B3*). I will use this example to explain a scenario that is also sometimes known as Mendel's Second Law.

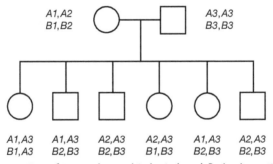

**Figure 3.4.** The segregation of two polymorphic loci, *A* and *B*. A schematic of a hypothetical family in which the inheritance of the mother's alleles at *A* (*A1* or *A2*) occurs independently of the inheritance of her alleles at *B* (*B1* or *B2*). The segregation of the alleles from each parent in their progeny is illustrated, as is the independent assortment of alleles from the two loci, *A* and *B*.

*Independent Assortment:* Independent assortment (sometimes referred to as Mendel's Second Law) describes when pairs of loci are inherited independently of one another, either because they reside on different chromosomes or because they are distant from each other on the same chromosome. In this new family, the inheritance of the mother's alleles at *A* (*A1* or *A2*) occurs independently of the inheritance of her alleles at *B* (*B1* or *B2*), because the number of children with each possible maternal allele combination (*A1 B1*, *A1 B2*, *A2 B2*, and *A2 B1*) occur at roughly equal frequency. Strong deviations from this pattern of marker allele inheritance, which reflects independent assortment, would be evidence of linkage between loci *A* and *B*.

In the next illustration (Fig. 3.5), I have added data for another polymorphic locus, *C*, to this same family. This time, I have replaced the noninformative data from the father with dashes, again to aid exposition. In this scenario, the pattern of inheritance we see in the offspring deviates from that expected from independent assortment: Whereas there is no correlation between the inheritance of the mother's *A* and *B* alleles, there is clearly some correlation between the inheritance of her *A* and *C* alleles. All the children who inherited *A1* also inherited *C1*, and all the children who inherited *A2* also inherited *C2*. Clearly, there is a correlation here in the inheritance of these loci, which implies linkage.

At this point, it is important that I highlight an ambiguity in the way in which DNA polymorphisms are detected. The fantastic methods we currently have in hand allow us to learn, for each individual, the presence or absence of particular alleles, but not their arrangement on the chromosomes (their so-called "linkage phase").

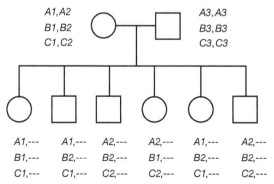

**Figure 3.5.** The segregation of three polymorphic loci originating from one parent. A schematic of a hypothetical family in which three loci (*A*, *B*, and *C*) are segregating in the mother and are homozygous in the father. Linkage (lack of recombination) between the alleles at loci *A* and *C* is shown: *A1* and *C1* are inherited together, as are *A2* and *C2*.

So, in the case of loci $A$ and $C$, there are two possibilities: $A1$ and $C1$ are located on one of the two homologous chromosomes and the alleles $A2$ and $C2$ are located on the other, or alternatively, $A1$ and $C2$ are on the same homolog and $A2$ and $C1$ are on the other. Clearly, the former arrangement is more likely, given the results in the illustration: If we assume the first arrangement, all the children are nonrecombinants. If we choose the alternative, all the children must be recombinants. I discuss linkage phase in more detail below.

**Linkage Phase:** As mentioned above, linkage phase refers to how the polymorphic loci mapped in a cross are physically arranged on their respective chromosomes. So, where alleles reside on the same homolog (i.e., when they are arranged *cis* to each other, or, as described in the classical literature, are in "coupling"), they will be inherited together in the absence of recombination. Where alleles are arranged *trans* to each other (or, as in the classical literature, are in "repulsion"), they will be segregated away from one another.

In the above example, the two linkage phases would be denoted as

$$\text{Phase 1 } (cis), \quad \frac{A1 \;\; C1}{A2 \;\; C2}; \qquad \text{Phase 2 } (trans), \quad \frac{A1 \;\; C2}{A2 \;\; C1}.$$

The data from the example illustrated in Figure 3.5 strongly support a complex hypothesis concerning the inheritance of these loci. In this hypothesis, linkage exists between loci $A$ and $C$, and they are therefore in phase 1 (*cis*) in the mother. This strongly suggests that one of the mother's parents passed on this same arrangement of alleles to her. Figure 3.6 illustrates the assignment of recombinants based on the assumption of phase.

To see this quantitatively, consider the possibility that the frequency of recombination $F_{rec}$ between loci $A$ and $C$ is 0.1 (which is equivalent to 10 cM, corresponding, in humans, to around 10 million bp). Then the probability that the parents in Figure 3.5 had six recombinant children would be 0.1 to the sixth power, or one in a million ($10^{-6}$). Thus, the data strongly support a *joint hypothesis*: that loci $A$ and $C$ are linked and that they are in a *phase 1* arrangement in the mother's genome. In this arrangement, loci $A$ and $C$ are therefore most likely to be inherited together. Clearly, this idea can be extended to more than two loci, provided they are all linked. Thus, alleles inherited together in *cis* will tend to be transmitted together.

**Haplotype:** This word describes the case in which the alleles of several linked polymorphic loci are inherited together. For example, in Figures 3.5 and 3.6, the loci $A1$ $C1$ (and also $A2$ $C2$) together form a haplotype. As we will see in Chapter 14, analyses of SNPs in human populations have revealed that some human haplotypes are

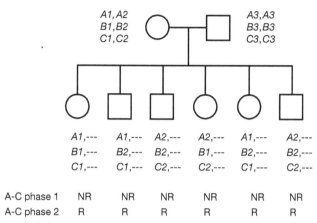

**Figure 3.6.** Assigning linkage phage. A schematic of marker allele inheritance in the progeny of the family illustrated in Figure 3.5. Here the progeny are assigned as being recombinant (R) or nonrecombinant (NR), based on assumptions made about linkage phase in the mother. In phase 1, alleles *A1* and *C1* (also *A2* and *C2*) are assumed to be in *cis*, and in phase 2, they are assumed to be in *trans* such that *A1* and *C2* (also *A2* and *C1*) are inherited together.

very ancient, with alleles having remained together and not separated by recombination for very many generations, back to the earliest humans.

## Genetics and Statistics

The sciences of genetics and statistics progressed together through the 20th century. I mentioned that one of Mendel's insights was that he needed to characterize and count very large numbers of plants in order to distinguish different ratios of progeny types. This was a big innovation, because in his time the use of statistical reasoning was uncommon, especially in biology. In recombination mapping, a major issue was the precision and reliability of fundamentally probabilistic determinations of genetic distance. In the early 1900s, statistical theory began to be applied to these issues; new theory was sometimes required and ultimately developed. This resulted in a very close connection in the academic world among geneticists, statisticians, and evolutionists, especially in the United Kingdom. Exemplary of this close connection was the work of Ronald Fisher, who made fundamental contributions to statistical, genetic, and evolutionary theory. He introduced many of the basic statistical methods still in use today.

The joint statistical testing of phase and linkage, taken together as a single hypothesis, is characteristic of human genetics. The basic methods of inference were introduced by Fisher, specifically as an approach to genetic measurements, in the 1930s. Much of the apparently complex mathematics and computing used today in human gene mapping comes down to comparing quantitatively such joint hypotheses. In the example I provide in Figure 3.6, the joint hypothesis is close linkage (actually, zero recombination) between loci $A$ and $C$ combined with phase 1 linkage of the alleles. The alternative hypothesis is that of independent assortment, in which each possible arrangement of alleles in the progeny is equally probable, as we see is the case for loci $A$ and $B$. Fisher's method of maximum likelihood provided the mathematical framework for evaluating such a hypothesis given a set of data from families in which genes of interest are segregating.

***LOD Score and Likelihood Ratio:*** Fisher's maximum likelihood method computes the probability of the hypothesis given the data in hand (e.g., the genotypes of the children in Figs. 3.5 and 3.6) and compares this to the probability that these data were the result of independent assortment (i.e., all the genotypes are equally probable). This calculation is carried out for all values of $\theta$, the frequency of recombination. For phase 1, there are no recombinant children, so the best value of $\theta$ is 0 for loci $A$ and $C$. The maximum likelihood method compares this value to the probability of random inheritance with no linkage (i.e., $\theta = 0.5$). The ratio of these calculated values, called the "likelihood ratio," is then used to evaluate the evidence for linkage.

It turns out that a reliable threshold for inferring linkage between loci, based on both theory and experimental evidence, is a likelihood ratio of 1000. It has become conventional to report the logarithm of this ratio, called the LOD score (for logarithm of the odds). A LOD score of 3 has long been the standard for determining linkage in human families, and few false linkages have been found using this threshold in family studies of simple Mendelian traits. Today, many thousands of genes that cause disease have been mapped by recombination mapping using DNA polymorphisms and likelihood statistics. In most of these cases, the actual causal genes have been isolated, and much has been learned about these diseases that could not have been in any other way.

For example, before human gene mapping was possible, it was clear that the childhood disease cystic fibrosis segregated in families as a simple recessive Mendelian factor. We understood very little about this disease until its causative gene was located, by linkage mapping in families, to human chromosome 7. This led to its isolation and subsequent sequencing, which revealed that cystic fibrosis is caused by a defect in the handling of chloride ions in various tissues.

However, family studies have also revealed the existence of traits and diseases for which no single Mendelian locus appears to be responsible. These reflect complexity in the mechanism of inheritance (which, as I discuss below, should not be confused with phenotypic complexity).

Even the most straightforward of disease-mapping cases, in which pheno-types caused by a disease map to a single locus segregating as a simple Mendelian trait, can be very complicated, because human physiology is complicated. Mapping the causative gene to a single locus means that these complicated pheno-types can all be traced back to a single locus, in some cases to just a single DNA mutation. Examples include cystic fibrosis, which is caused by defects in a gene that encodes a transporter of chloride ions, or Lesch–Nyhan syndrome, which is caused by defects in a metabolic pathway that contributes to the biosyn-thesis of purine nucleotides. The cystic fibrosis phenotype includes severe effects on lung function (including susceptibility to persistent infection), as well as on digestive function. And among the phenotypic hallmarks of Lesch–Nyhan syn-drome are characteristic self-mutilation behaviors that could, on the surface, be thought of as a complicated neurological defect.

Conversely, deceptively simple inherited disease phenotypes, such as high blood pressure or cholesterol levels, turn out to involve changes at many different loci. It is this situation that we describe as "complex inheritance." This is an area of very great importance to human health, and I return to it in Chapter 14.

## INTRODUCTORY BIOGRAPHIES

**Thomas Hunt Morgan (1866–1945)** was an American geneticist, embryologist, and evolutionary biologist. He used *Drosophila melanogaster* to establish the chromo-some theory of heredity and established the famous "fly room" at Columbia Uni-versity that collectively established *Drosophila* as the premier model organism for experimental genetics. He founded the Division of Genetics at Caltech, which became the center of the emerging field of molecular biology.

**Alfred Sturtevant (1891–1970)** was an American geneticist who, while still an undergraduate working with Thomas Hunt Morgan at Columbia University, con-ceived the idea of quantitative recombination-based maps of a chromosome. He was also a pioneer in developmental biology, devising a quantitative method for following the spatial relationships of developing organs in flies.

**J.B.S. Haldane (1892–1964)** was a Scottish-born physiologist, mathematician, and geneticist. He contributed in a major way to the theory of enzyme kinetics and was a pioneer of genetic and evolution theory, especially population genetics. He intro-duced many of the ideas of the "modern synthesis" that reconciled Mendelian inheritance with the observations of quantitative traits. It was Haldane who first introduced the nomenclature *cis* and *trans* in 1941 to replace the classical "coupling" and "repulsion," respectively, to describe linkage phase.

**Ronald Fisher (1890–1962)** was a British statistician, geneticist, and evolutionary biologist. He was a founder of modern statistics, which he developed for population genetics evolutionary theory. His statistical methods, including analysis of variance (ANOVA) and the method of maximum likelihood, have become the centerpieces of statistical analysis in virtually every field of science.

CHAPTER 4

# Pathway Analysis

## Metabolic Pathways

All of life is basically chemistry. All organisms are built by a vast number of hierarchically and temporally ordered chemical reactions, most of which are catalyzed by proteins known as enzymes. These reactions are generally referred to as "biochemical" reactions, and the metabolism of a cell is the ensemble of its biochemical reactions. The energy needed for life is provided via a series of biochemical reactions that result in the production of high-energy molecules; these molecules drive chemical transformations that would otherwise be energetically or kinetically unfavorable. The production of these high-energy molecules (principally adenosine triphosphate [ATP], but there are others) requires a continuing source of energy. Humans live mainly by aerobic metabolism, in which the oxidation of sugar (the ultimate source of which is plant metabolism, driven directly by energy from the sun) is broken down into small, necessarily sequential, chemical steps, some of which enable the capture of free energy. This free energy is ultimately collected in the form of the high-energy phosphodiester chemical bonds that are present in ATP.

The cells that make up our bodies are also built by biochemical reactions, which are controlled ultimately by instructions encoded in our DNA. Many of these biochemical reactions are, once again, necessarily hierarchically and temporally ordered. For example, to produce proteins, we need not only the information encoded in our DNA but also a continuing supply of the 20 amino acid monomers that are polymerized to form proteins. To generate amino acids and the nucleotides that form DNA, a series of ordered biochemical reactions are required to transform the intermediates of metabolism into these components. Each series of metabolic reactions can be thought of as a "pathway"; each pathway begins with a particular starting metabolite that arises from central, energy-creating metabolism and ends with one of the amino acids. Thus, cellular metabolism consists of a vast network of metabolic pathways.

*Pathway:* This word is a metaphor that has become embedded in our thinking about metabolism, and about other biological processes as well, such as the networks that transduce biological signals. It is used to describe biological events that must occur via an ordered series of steps.

*Upstream and Downstream:* These popular metaphors represent the directionality of the flow of material through a pathway. They refer to the position of a particular intermediate, with respect to the start or end of a pathway; upstream reflects events that take place at or toward the beginning of a pathway, and downstream, those that take place on the way to a pathway's end product. The reader should be aware that with respect to signaling pathways, these metaphors refer to the flow of information rather than the flow of chemical material per se. In such cases, as we shall shortly see, some subtleties and even potential contradictions can emerge in the diverse uses of these terms.

Because virtually every biochemical reaction requires catalysis by enzymes, the amino acid sequence of which is specified by DNA sequence, the loss of function of many genes can be recognized as an alteration (usually a loss) in the ability to execute a specific biochemical reaction. Such a loss of a specific chemical reaction has some predictable consequences: Molecules that are downstream of the missing reaction (i.e., downstream intermediates) will no longer be made, but molecules that are upstream (i.e., intermediates upstream of the missing reaction) will be present and, more often than not, will accumulate to levels not seen in cells with an intact pathway, like water upstream of a dam in a river.

Historically, the genetic analysis of biochemical pathways has been entwined with the discovery of the pathways themselves. Relatively few metabolic pathways were deduced by chemistry alone. The longer and more complex series of metabolic chemical reactions were elucidated using information about the identities of pathway intermediates. These were discovered from the analysis of molecules that had accumulated in cells in which a genetic mutation had blocked part of the pathway. Commonly, these discoveries were made in standard model microbial species favored by experimentalists, such as the bacterium *Escherichia coli* and the yeast *Saccharomyces cerevisiae*, both of which grow on very simple mixtures of sugar and inorganic nutrients, as explained below.

*Prototroph:* Many microorganisms, notably *E. coli* and *S. cerevisiae*, are prototrophs—meaning that they can synthesize all their amino acids, nucleic acids, and other cellular constituents from inorganic nutrients. This allows them to be grown in simple media, consisting only of sources of carbon (C), nitrogen (N), sulfur (S), phosphorus (P), and some salts. Most of the metabolic pathways that produce amino acids and nucleotides were worked out using mutants of such organisms to identify the chemical intermediates along the pathway and to isolate, purify, and characterize the enzymes that carry out these reactions. By contrast, an auxotroph is a microorganism that is unable to synthesize a particular compound required for its growth.

*Auxotrophic Mutations:* This type of mutation (often a loss of function) usually occurs in a microbial gene that encodes a product required for the synthesis of an essential metabolite, such as an amino acid or one of the bases that make up DNA and/or RNA. The phenotype of an auxotroph is a requirement for its culture medium to be supplemented with one or more specific nutrients (typically the end products of the pathway). If, and only if, the nutrient is provided, the auxotroph will grow; in contrast, a prototroph will not require the added nutrient.

Mutant cells can also be chemically different in other ways. They can fail to make not only a product of a pathway but all the intermediates between the blocked step and the product as well. In the example illustrated in Figure 4.1, the phenotype of a null mutation in the gene encoding enzyme E2 will include the failure to make intermediates $I_2$, $I_3$, and $I_4$, as well as the product P. When starved of the product (P), the E2 mutant will be a good source of the chemical intermediate immediately before the block ($I_1$), which typically is overproduced and accumulates as the regulatory systems of the cell try to compensate for the deficiency in the product of the pathway. The regulation of metabolic pathways is a subject we return to in Chapter 5.

Today, one might analyze these additional phenotypes directly with a mass spectrometer, a device used to detect the type and amount of chemicals present in a sample. The early microbiologists who worked out the basic metabolic pathways of microorganisms devised a more subtle and quintessentially biological approach: They collected many auxotrophic mutations with similar growth requirements and selected those that fell into different complementation groups.

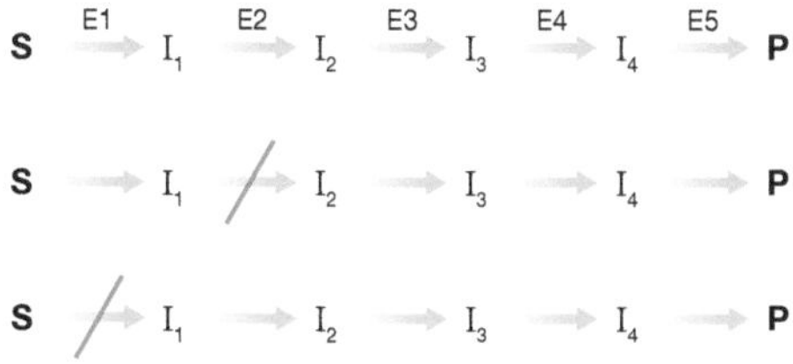

Figure 4.1. A hypothetical metabolic pathway. A schematic of a hypothetical metabolic pathway, which begins at a substrate (S) and proceeds through sequential intermediates ($I_1$, $I_2$, $I_3$, and $I_4$) to produce the end product (P). The enzymes E1–E5 catalyze the chemical transformations as shown. The vertical lines at E1 and E2 denote null mutations that result in the loss of enzyme function.

They then determined whether any of them could cross-feed any of the others, as I explain below.

**Cross-Feeding:** Cross-feeding is a historically important biological assay for identifying those metabolic intermediates that can be taken up by cells (not all compounds cross membranes easily). In the example given in Figure 4.1, a mutant blocked at E2 should be able to feed another mutant blocked at E1, because it can produce the first missing intermediate ($I_1$) of this pathway, which is lost in the E2 mutant; it cannot feed mutants blocked in any of the other steps of the pathway. Both the identity of the source (in this case the mutant in E2) and the quantitative assay (based on the ability of E1 mutants to grow) facilitate the isolation of this pathway's intermediates and the purification of the enzymes that transform one intermediate into the next.

To summarize:

> ▸ *A null mutant that has a blocked metabolic or biosynthetic pathway is a **source** of chemical intermediates that are synthesized before the blocked step and can be fed by chemical intermediates synthesized after the block.*

## Using Double Mutants to Discern the Order of Steps in Metabolic Pathways

As suggested above, the order of steps in a metabolic or biosynthetic pathway can be deduced from the properties of double mutants in that pathway. This is true even when there is limited information about the underlying biochemistry. The approach used to assess where a gene functions in a pathway rests, once again, on complementation and recombination, which are best thought of in terms of an implicit experiment.

To analyze a double mutant, the phenotypes of the mutations at the different steps must be distinguishable in some way. This can become complicated, however, when there is more than one phenotype to assess. As in the assessment of dominance, a single mutation can have different phenotypes that lead to different conclusions in an implicit experiment.

In the hypothetical pathway discussed above (Fig. 4.1), the phenotypes of the single mutants in the E2 gene and E4 gene differ in two main ways. In one mutant (in this example, E4), its phenotype is the ability to serve as a *donor* in a cross-feeding experiment, which allows one to determine the presence or absence of an intermediate. This phenotype could also be assessed by direct chemical assay, for instance, with a mass spectrometer, which would also identify an accumulated intermediate. For the other mutant (in this example, E2), the phenotype is the ability to be fed as a *recipient* in a cross-feeding experiment.

The donor phenotype follows the flow of chemical material in the pathway—each mutant makes all of the upstream intermediates that are generated before the block imposed by the absence of a gene product; it makes none of the downstream intermediates beyond this block. In Figure 4.2, we see that the phenotype of the double mutant, which lacks both the E2 and E4 enzymes, resembles the E2 mutant with respect to its ability to make the intermediates in this pathway. From the viewpoint of the donor phenotype (and of direct chemical analysis), the E2 block is *upstream* of the E4 block, and both E2 and the double mutant make only the first intermediate, $I_1$. Conversely, from the viewpoint of the cross-feeding recipient phenotype (i.e., the ability to be fed by pathway intermediates), the E2/E4 double mutant resembles the E4 single mutant (and *not* the E2 mutant) with respect to its ability to grow when fed pathway intermediates $I_2$, $I_3$, or $I_4$. Thus, a purely genetic assessment of "upstream" and "downstream" is not possible without a clear understanding of the relationship of the mutant phenotypes to the pathway.

The reader should recognize that the pathway order recovered by either analysis is actually the same. The difference is only in the way we classify mutants with respect to the flow of material through the pathway. The origin of this difference lies with the fact that a block in a pathway produces two outcomes: a loss of the intermediates beyond the block *and* a requirement for these same intermediates in a cross-feeding experiment. For biosynthetic pathways, one always wants to use the source (or chemical analysis) phenotype, so that the analysis follows

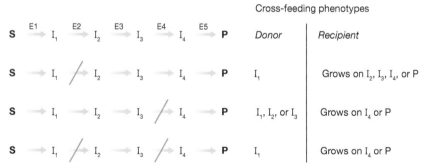

**Figure 4.2.** An illustration of cross-feeding experiments with metabolic mutants. Donor and recipient phenotypes (see text) for each mutant are shown on the *right*. Note that the double mutant E2/E4 shares the same donor phenotype as the E2 single mutant and the same recipient phenotype as the E4 single mutant. The vertical lines at E2 and E4 denote mutations that result in the loss of enzyme function. E, enzyme; I, intermediate; P, product; S, substrate.

the flow of material through a pathway. Double mutants in the pathway will then always resemble the upstream single mutants.

The reader may recall encountering a similar dependence of a genetic inference on the phenotype assessed in the discussion of dominance in Chapter 2. The key to avoiding confusion is a clear understanding of the mutant phenotypes in relation to the pathway under study. In the case of metabolic and assembly pathways, this is rarely a problem; one just follows the flow of material, as indicated above. However, the analysis of double mutants in regulatory and signal transduction pathways, in which the process is information, as opposed to a series of chemical transformations or steps in building a structure, requires particular care. I will return to these pathways later in this chapter.

## Morphogenesis and Assembly Pathways

Numerous cellular structures are assemblies of many different proteins that come together in an ordered series of steps, which include a surprising number of stable intermediate structures. Many of the pathways that lead to the biosynthesis and assembly of complex and essential cellular structures have been revealed by the analysis of genetic pathways. These analyses have much in common with the way in which metabolic pathways were worked out.

At first, most biologists imagined that these structures formed as a result of pathways that consist of an ordered series of enzymatic functions, analogous to the metabolic pathways that generate small molecules. However, genetic analyses, using mutants to identify and characterize intermediate structures, quickly showed that most of the pathways that generate cellular structures are examples of the self-assembly of proteins, in which relatively stable intermediate structures serve as a substrate for the next step in a structure's assembly.

The paradigm example of this is the dissection of the pathways that bring about the morphogenesis of a bacteriophage called T4, which was carried out by the bacteriophage geneticists Robert Edgar and William Wood in the 1960s. Bacteriophages (commonly referred to as "phages") are viruses that grow on bacteria. The mutations Edgar and Wood used are called conditional-lethal mutations, and I will have much more to say about both phages and conditional-lethal mutations in Chapter 6. For the purpose of understanding how they were used to study morphogenetic and assembly pathways, all we really need to know at this stage is that there are several ways in which a change in DNA can result in the loss of a genetically specified function in one environment but not in another. Under one set of conditions, called "permissive," the organism is unimpaired, but under another,

called "nonpermissive" (or "restrictive"), the organism is missing the function specified by the functional gene in which the mutation lies.

The conceptually simplest conditional-lethal mutations are those that cause
the gene product to be unstable or missing at an elevated temperature. These
"temperature-sensitive" (*ts*) mutations allow us to study loss-of-function mutations that are lethal to an organism by growing it at a low (permissive) temperature, which allows the affected gene product to function; we then simply move it
to a higher, nonpermissive temperature, at which the gene product cannot function or is absent, and the organism cannot grow. The reader has probably realized
already that auxotrophs exemplify another class of conditional-lethal mutants:
ones that cannot grow in media lacking a particular metabolite but that can
grow if the metabolite is supplied.

By 1963, a comprehensive set of conditional-lethal phage T4 mutants had
been collected by Edgar and his colleagues.[1] They classified these into more
than 50 functional genes by complementation tests. As we will see in Chapter
6, complementation and recombination tests can readily be done using viruses,
and their interpretation in terms of biological function is the same as it is for other
kinds of organisms.

The morphological phenotypes of these mutants, representing mutations in
dozens of different genes, were determined by electron microscopy, because phage
T4, like most viruses, is very small, much smaller even than a bacterial cell. These
microscopy studies identified numerous distinguishable morphological phenotypes, including varying forms of incomplete heads, tails, and tail fibers (see Fig.
4.3). Using just these phenotypes, Edgar and colleagues obtained information about
the order in which these structures were assembled from double mutants, deriving a
sequence of events, just as in a metabolic pathway, by assessing the ability of
mutants to produce different products (in this case phage heads, tails, or tail fibers).

The great advance in understanding these pathways came from the discovery
that the assembly of viable phages will occur in a mixture of cell-free extracts made from cells that had been infected under nonpermissive conditions
with two different conditional-lethal mutants, each of which by itself produces essentially no viable particles. Edgar and Wood called this assay "in vitro
complementation."

With this assay and their many mutants, Edgar and Wood found, using a logic
that is very similar to the donor logic employed in metabolic pathway analysis,

---

[1] Epstein RH, Bolle A, Steinberg CM, Kellenberger E, Boy de la Tour E, Chevalley R, Edgar RS,
Susman M, Denhardt G, Lielusis A. 1963. Physiological studies of conditional lethal mutants of
bacteriophage T4D. *Cold Spring Harb Symp Quant Biol* **28:** 375–394.

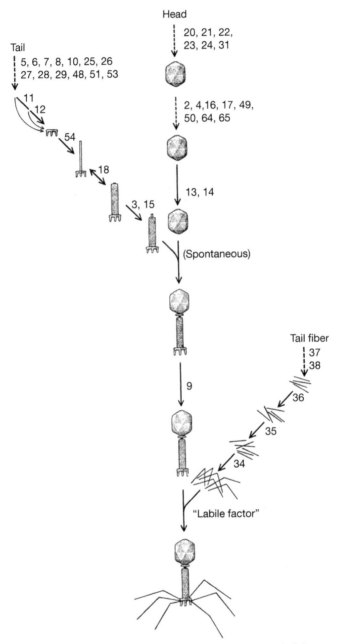

**Figure 4.3.** The assembly pathway of bacteriophage T4. An illustration of a fully assembled T4 bacteriophage (bottom), together with the assembly pathways for its head, tail, and tail fibers. Mutants in these pathways provide distinguishable morphological phenotypes for use in complementation tests. The numbers are the names of the T4 genes (complementation groups). (Reproduced from Wood WR, Edgar RS. 1967. *Sci Am 217:* 61–66, with permission from the estate of Bunji Tagawa.)

that the virus-assembly process is divided into three independent pathways.[2] One pathway produces the tail, another produces the tail fibers, and a third produces the head, which is already filled with DNA. Each of the nonpermissive mutant extracts defective in producing tails could act as donors for head assembly when mixed with extracts of mutants defective in producing heads; each of the mutants defective in producing heads was a tail donor; and so on.

There are some key lessons to be drawn from this assembly pathway that apply both to other phage assembly pathways and to the vast numbers of such pathways that build subcellular macromolecular structures. Probably the most important is that intermediate biological structures assembled from protein and/or RNA subunits tend to be stable, and remain so until the conditions for the next step are met. Second, the donor logic derived from cross-feeding works for assembly pathways just as it does for metabolic pathways. In both cases, the flow of material defines the directionality of the pathway and predicts the order in which genes function in that pathway.

## Regulatory and Signal Transduction Pathways

The use of the pathway metaphor is not limited to metabolic and assembly pathways. It is also used to describe sequential gene activity during a physiological or genetic response to environmental change (signal transduction pathways) and to describe the decisions made among alternative cell fates (developmental regulatory pathways). Examples of such pathways include the processes that determine sex in flies, worms, and humans (each of these is different, but all have in common two alternative outcomes: male and female); the processes that activate the immune system when a foreign antigen (such as a virus) is detected; and the processes that determine whether cells divide or remain quiescent.

The salient property of these pathways is that they reflect the transfer of information that ultimately determines which of two binary outcomes will occur. Although they involve the serial activity of gene products (usually proteins), they differ from metabolic and regulatory pathways in that they describe a decision process rather than a series of chemical transformations or assembly steps. In this way, the regulatory and signal transduction pathways resemble electronic switching systems. The genetic consequences of null mutations in these pathways, as well the consequences of double mutations in them, are consequently different from those in metabolic or assembly pathways.

---

[2] Wood WB, Edgar RS. 1967. Building a bacterial virus. *Sci Am* **217**: 60–74.

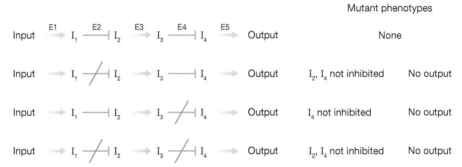

**Figure 4.4.** A hypothetical regulatory or signal transduction pathway. The input into this type of pathway is typically a ligand-binding event that involves E1, producing $I_1$ (which could be a modified E1, or even a small molecule). $I_1$ inhibits the formation of $I_2$, an event that involves E2; $I_2$ inhibition stimulates the formation of $I_3$, which inhibits the formation of $I_4$, whose absence is essential for the output. The slanted vertical lines denote mutations that result in the loss of enzyme function. E, enzyme or signaling molecule; I, intermediate.

Figure 4.4 shows a hypothetical pathway that carries information that ultimately results in an output, such as the determination of the male sex, the activation of an immune response, or cell division. Here, the intermediates ($I_1$–$I_5$) are not necessarily molecules; for example, $I_1$ could be the ligand-bound form of a receptor for a sex hormone (the product of the *E1* gene). Some of the intermediates could be small molecules; others could be proteins modified by the addition of a phosphoryl or methyl group. The main point is that, in this scenario, the phenotype of double mutants is the same as that of single mutants present at *later* steps in this information pathway. In this way, regulatory and signal transduction pathways resemble metabolic pathways that are phenotypically assessed as *recipients* in a cross-feeding experiment (see Fig. 4.2). In both, the double mutants resemble mutations in the later step of the pathway (for example, the E4/E2 double mutant shown in Fig. 4.2).

Therefore, it is vital to understand the nature of the pathway under genetic investigation by the method of double-mutant analysis, as well as by the phenotypic assessment being used.

▸ *When material is being transformed or structures are being built, and the assessment is the completion of the process, the phenotype of double mutants resembles that of the earlier of the two single, component mutations.*

▸ *When information determines a decision in cell fate or outcome, and the phenotypic assessment is the outcome itself, the phenotype resembles that of the later of the two single, component mutations.*

*Epistasis:* In the original definition of this problematic word, about which I say more in Chapter 10, one phenotype is said to be "epistatic" to another when the phenotype of one masks the phenotype of the other. So, in the hypothetical metabolic or biosynthetic pathway illustrated in Figure 4.2, the phenotype of E2 is epistatic to that of E4 because the double mutant has the E2 phenotype. The early English geneticist William Bateson (already encountered in Chapter 1) introduced this word, and provided the idea of one phenotype "masking" another. In the metabolic pathway illustrated in Figure 4.2, the E4 mutant phenotype is "masked" by the presence of the E2 mutation in the double mutant. In the hypothetical signal transduction pathway illustrated in Figure 4.4, the E2 mutant phenotype is masked by the presence of the E4 mutation in the double mutant. In Figure 4.2, a mutation in E2 is said to be epistatic to the mutation in E4; in Figure 4.4, the mutation in E4 is said to be epistatic to the mutation in E2. This use of double mutants is what experimental geneticists call, even today, a "test of epistasis."

The reader may well find even this use of the word "epistasis" confusing. Most students of genetics I have encountered in my teaching career have been confused by this word. As I explain in Chapter 10, I strongly recommend avoiding the use of this word, mainly because of the confusion caused by its being given another meaning in quantitative genetics. Nevertheless, readers should be aware that this word is regularly used to describe double-mutant analysis and is generally understood by experimental geneticists in both the context of metabolism and biosynthesis, on the one hand, and in regulation and signal transduction, on the other. In either case, remembering Bateson's idea of "masking" makes it more intuitive as to which gene is epistatic to which.

## INTRODUCTORY BIOGRAPHIES

**Robert S. Edgar (b. 1930)** is an American geneticist and molecular biologist who laid the foundations of bacteriophage genetics in his studies of phage T4. He developed conditional lethals and used them to map the genome of T4 and to survey the functions of the essential T4 genes. With William Wood, he worked out the pathways of T4 phage morphogenesis.

**William B. Wood (b. 1938)** is an American molecular biologist and biochemist who, with Robert Edgar, worked out the pathways of T4 morphogenesis. He also was a pioneer in establishing the nematode *Caenorhabditis elegans* as a model for studying the molecular biology of development.

CHAPTER 5

# Regulation of Metabolic Pathways

## Introduction

To reproduce, all cells must synthesize hundreds of different small molecules containing carbon, nitrogen, sulfur, and phosphorus. Animals, fungi, and most bacteria make these molecules, starting with glucose as the source of carbon and energy, and using different metabolic pathways that lead to each of the small-molecule end products. The network of metabolic pathways that lead from glucose to the many end products required for life is necessarily complicated, as illustrated in Figure 5.1, which summarizes part of a consensus network of metabolic pathways of many organisms, including bacteria, plants, and animals. A cell has to have exactly the right concentrations of each of the chemicals in the network at each stage of its life cycle. This is an engineering problem that is at least as complicated as arranging for every airport in the world to have every plane arrive and depart on time, with a full load of passengers, regardless of the weather. Like air traffic, the system must deal with local perturbations that can affect distant nodes of the network. For example, a weather delay in Chicago causes ripples of delay in many other cities that require constant adjustments to be made to gate and runway availability, to say nothing of the issues around air traffic control to prevent airplanes from flying into one another.

The evolutionary advantage of efficiency in metabolism is as strong as it is obvious. A metabolically inefficient organism will be at a severe reproductive disadvantage relative to one that, like a Toyota factory, organizes its supply of parts on a "just-in-time" basis. This strategy minimizes the inefficiencies of shortages and oversupply, and thereby also limits requirement for storage, materials, and energy. So it comes as no surprise to learn that metabolic pathways, and indeed all biological pathways, are highly regulated. The regulatory mechanisms that bring about this regulation, on the one hand, serve to avoid shortages of essential intermediates, and on the other, minimize the waste of material and energy that is inherent in making more of any chemical entity than is required.

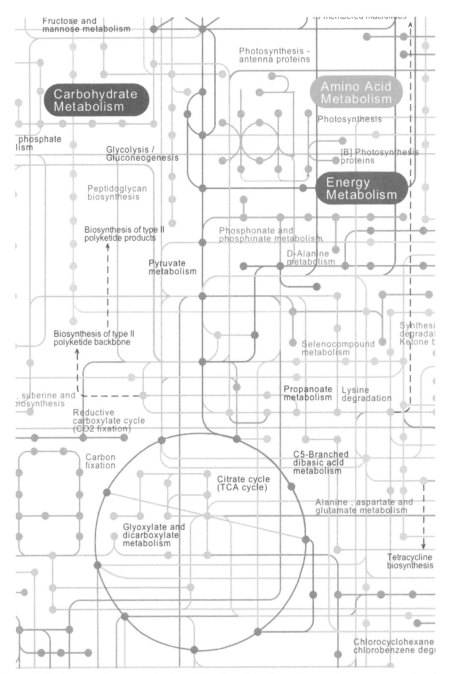

**Figure 5.1.** A part of a general schematic of metabolic pathways. A consensus network of the metabolic pathways of many organisms, including bacteria, plants, and animals. (Image reproduced, with permission, from Kyoto Encyclopedia of Genes and Genomes; http://www .genome.jp/kegg/.)

For cells to grow efficiently, each pathway should have a flux of molecules passing through it that matches an organism's requirements, resulting in what physiologists call "balanced growth." This is the cellular equivalent of Toyota's "just-in-time" system. We know, experimentally, that organisms grow efficiently in a balanced way because they do not produce more metabolites than they need. For example, a culture of prototrophic bacteria growing in a simple medium of salts and glucose does not excrete amino acids. The ability of an organism to regulate its metabolism in this way holds true for all circumstances that are consistent with viability and growth.

Twenty amino acids are required for protein synthesis; they are supplied to the cellular machinery, which produces proteins in a coordinated fashion, such that just the right amounts are available when needed. The metabolic pathways that produce amino acids begin with just a few types of molecules, which are produced in the process of extracting energy from sugars. This means that often several pathways begin with the same starting molecules, and they must avoid interfering with one another to ensure that one metabolic pathway does not steal substrates or energy (in the form of high-energy ATP molecules) from another. Some amino acids are themselves precursors of other molecules that are required by the cell for purposes other than protein synthesis. In such cases, the cell must partition the supply of such an amino acid between the ribosomes, which consume most of a cell's amino acids in the production of proteins, and the metabolic pathways that use it to produce other necessary chemicals.

A nice illustration of this regulated balance between supply and demand is the synthesis of neurotransmitters, the chemicals that carry signals from one nerve cell to another across a synapse. Establishing the correct levels of neurotransmitters is essential for neurological function. Most neurotransmitters are simple chemical derivatives of amino acids: For example, dopamine and adrenaline are derivatives of the amino acid phenylalanine; serotonin and melatonin are derivatives of the amino acid tryptophan. In some instances, the neurotransmitters are the amino acids themselves, as is the case with aspartate and glutamate. Neurons, cells that have to both make protein and produce specific neurotransmitters, require regulatory systems that keep these pathways in balance, lest the cell fail in one or the other of these essential processes.

The most basic and intuitive regulatory systems are feedback mechanisms. These can work at the level of enzyme activity or at the level of their biosynthesis. Both levels of regulation appear to have evolved to enable a cell to achieve balanced growth under a great variety of environmental conditions. Below, I discuss what we mean by the word "regulation."

*Regulation:* Geneticists use the word "regulation" in a way that is significantly less broad than its many common usages in the English language (the *Oxford*

*English Dictionary* gives several alternatives for the verb "to regulate"). Like most geneticists, I limit my use of "regulation" to mean "to control, modify, or adjust with reference to some principle, standard, or norm; to alter in response to a situation, set of circumstances, etc." Some published textbooks and articles use the word too broadly, in my opinion. For example, I would not say that a gene "regulates" its product simply because it encodes its amino acid sequence. I restrict my use of the word to situations in which the expression or function of a molecule is "controlled, modified, or adjusted" in response to changes in circumstances, usually the internal or external environment in which a cell finds itself. It is this response that is the hallmark of "regulation" in biology.

## Regulation at the Level of Enzyme Activity

As I mentioned above, when a metabolic pathway is blocked by a mutation, the molecule that is the substrate of the missing enzyme tends to accumulate. The pioneers of modern biochemistry observed that this accumulation of intermediates is considerable only when the end product of the same pathway is limiting or absent. Over the years, this has turned out generally to be the case for metabolism: When the end product is plentiful, there is no accumulation of intermediates, even in mutants.

One way of studying this phenomenon is to let an auxotroph (let us say, for example, one that requires the amino acid tryptophan) grow in growth-limiting amounts of added tryptophan. In the 1950s, the American chemist Aaron Novick and the Hungarian physicist and inventor Leo Szilard invented the chemostat, a simple device that allows microorganisms to grow continuously under nutrient-limiting conditions. With this device, they showed that a chemical intermediate (called indole-glycerol-phosphate) accumulates in cultures of a tryptophan auxotroph of *E. coli*, the growth of which is limited by the external supply of tryptophan. When they then added more tryptophan to the culture medium, the accumulation of this intermediate stopped almost immediately, and no new precursor was made. Novick and Szilard hypothesized that the end product, tryptophan, exerts feedback control on its own metabolic pathway, stopping the production of metabolites at the early steps of the pathway. This makes sense, as it is tryptophan that is required for protein synthesis; when sufficient tryptophan is available, making more wastes both material and energy.

When Novick and Szilard allowed an auxotroph to starve after being cultured in adequate amounts of tryptophan, the intermediate began to accumulate just as the externally supplied tryptophan became depleted. In this environment, opening the gate to the pathway now makes sense, as the cell has reached the point of needing to produce more tryptophan.

Novick and Szilard were able also to show that this phenomenon did not involve the synthesis of new proteins; thus, the inhibition had to be exerted at the level of enzyme activity rather than at the level of synthesis. Many other pathways were soon found to behave in a similar fashion. And further work by other researchers showed that pathway inhibition by an end product could occur in cell-free extracts. These experiments confirmed that regulation is exerted directly at the level of enzyme activity.

*Feedback (or End-Product) Inhibition:* Feedback inhibition refers to the direct inhibition—by the end product of a metabolic pathway—of the enzymes that catalyze the first step in that pathway. The mechanism usually involves the end product binding to a site on that enzyme that is not involved in catalysis. For example, the first enzyme dedicated to tryptophan biosynthesis converts chorismic acid to anthranilic acid. It is feedback-inhibited by tryptophan. Even if the reader is, like the author, no organic chemist, it should be clear that chorismic acid and tryptophan are very unlike each other (Fig. 5.2).

*Allosteric Inhibition:* Allosteric inhibition is the most common mechanism for feedback inhibition. It derives from the French word *"allosterie"*—a term introduced by the renowned French microbiologist and geneticist Jacques Monod in 1961. The protein conformation of an allosteric enzyme changes when it is bound by an end product that is not a chemical analog of the substrate. The end product does not bind to the active site of the enzyme (the region that catalyzes reactions), but instead to another site, which is usually distant from the active site. When the concentration of the end product is high, the enzyme is inhibited because it is bound to the end product and its conformation is altered; when the concentration of the end product falls, the enzyme is no longer bound and it regains activity as it assumes its active conformation once again.

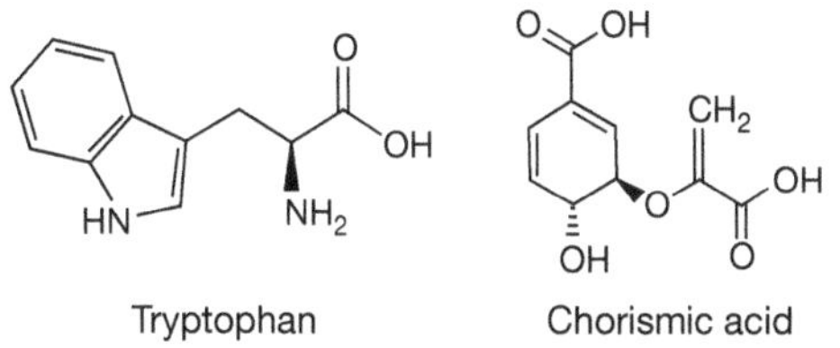

Tryptophan                Chorismic acid

Figure 5.2. Chemical structures of tryptophan and chorismic acid. *(Left)* Structure of the amino acid tryptophan, the end product of the tryptophan biosynthesis pathway. *(Right)* The structure of chorismic acid, an intermediate in tryptophan biosynthesis.

The consequences of allosteric interactions in proteins are not limited to inhibition. Many other kinds of regulatory changes are the results of such interactions.

*Allosteric Activation:* This refers to a situation in which an enzyme is bound by a ligand, which can be a small molecule (like an amino acid that is not a substrate or a product of the enzyme) or another protein, resulting in a change in conformation that causes the enzyme to be more active than it was before.

*Allosteric Modulation:* This newer, more general term is agnostic with respect to whether a conformational change induced by the binding of a ligand causes activation or inhibition.

*Posttranslational Protein Modification:* A very common class of change occurs to proteins after the polypeptide chain has been synthesized, called a post-translational modification. Common forms of such modifications can activate or inhibit enzyme activity and include phosphorylation (proteins do not normally contain phosphorus) and methylation, although there are many others. These chemical modifications are produced by the activity of specific enzymes and can be reversed by enzymes that remove the added phosphoryl or methyl groups.

These modifications, like simple allosteric interactions, enable the fast and reversible activation or inhibition of many kinds of biological pathways. In the late 1970s, the American biochemist Daniel Koshland elucidated what has become the paradigm example of regulation by a reversible protein modification. Koshland discovered that protein methylation is involved in the ability of bacteria to swim up (or down) the gradient of a chemical attractant (or repellent). As the concentration of the attractant changes, the sensitivity of the bacteria's chemotaxis machinery to that gradient adapts over a concentration range of many orders of magnitude. This reversible adaptation is the result of the addition or removal of methyl groups to the protein that binds the attractant (or repellent) chemical.

I have produced an example (Fig. 5.3) to illustrate the way in which metabolic coordination is achieved by feedback inhibition that is mediated by the end product of a pathway. In this example, the end product P inhibits enzyme E2, and the end product Q inhibits enzyme E6. Enzymes E2 and E6 are the points at which each pathway becomes dedicated to producing its particular end product. Their substrates also do not chemically resemble the end product (as is generally the case). Feedback inhibition serves to partition the common precursor $I_1$ between the P and Q pathways. This occurs because the first dedicated step of each

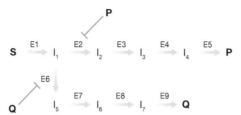

**Figure 5.3.** Feedback inhibition mediated by pathway end products. A hypothetical metabolic pathway that leads to the synthesis of two end products (P and Q) from the same precursor (S). Enzymes E2 and E6 are regulated by the concentrations of the end products P and Q, respectively. I, intermediates.

pathway will convert this precursor in proportions that are dictated by the levels of P and Q. This effectively provides on-demand flux control for each of the two pathways.

One might wonder how E1 might then be regulated, because in this model (Fig. 5.3), the concentration of $I_1$, the first intermediate common to both pathways, is not regulated. Often one finds in such cases two similar genes that encode two very similar enzymes that catalyze the same reaction (S→$I_1$). The enzymes differ in that one is allosterically inhibited by P and the other is allosterically inhibited by Q. The genes that specify these proteins are called "paralogs." Our example illustrates one way in which duplicated and then diverged genes, which are very abundant in higher organisms, might confer a selective advantage. I will return to discuss issues concerning duplicate genes and paralogs in Chapter 13.

## Regulation at the Level of Enzyme Synthesis

The same kind of experiment used by Novick and Szilard to demonstrate the phenomenon of end-product inhibition was also used to demonstrate that the abundance of end product regulates the *synthesis* of all the enzymes in a given pathway, as well feeding back on the *activity* of its earliest steps. In an auxotroph grown in limiting tryptophan, for example, all the enzymes of the tryptophan pathway are abundant and actively synthesized, even though the cell is starving for tryptophan because of an enzymatic defect. In contrast, cells grown in excess tryptophan appear to contain little or none of these enzymes. Furthermore, prototrophs that make their own tryptophan have low levels of the intermediate enzymes, much lower than those found in an auxotroph in limiting tryptophan. If a tryptophan auxotroph is allowed to starve for this nutrient, the robust synthesis of pathway enzymes begins just as the nutrient is exhausted. Unlike the accumulation of the pathway intermediate indole-glycerol-phosphate, the synthesis of pathway enzymes occurs relatively slowly because it requires active protein synthesis. It takes many minutes to make enzymes de novo from the instructions encoded in the DNA, which, as we shall see, is the level at which this regulation is exerted.

*Repression:* In this context, repression describes when a metabolite has a negative effect on the rate of protein biosynthesis, as in the case of tryptophan, where excess tryptophan results in a reduced rate of production of all the pathway's enzymes, whereas limiting tryptophan results in an increased rate.

Jacques Monod and his equally renowned colleague, the French biologist François Jacob, demonstrated another kind of regulation of protein abundance by a metabolite present in the growth medium of *E. coli*. They found that the enzymes that metabolize lactose (a sugar in milk) are present only when lactose is itself present in the culture medium. Jacob and Monod in 1961 published a truly magnificent series of experiments, which I return to in Chapter 11, which showed that the presence of lactose triggers the biosynthesis of the lactose-metabolizing enzymes. These enzymes are absent until lactose is added to the medium, and their appearance after its addition requires protein synthesis.

An evolutionary perspective here is helpful. When there is no lactose available, making the enzymes required for its utilization is an unnecessary expense for a cell. However, a bacterium that normally encounters lactose from time to time has an obvious evolutionary advantage if it can produce such enzymes and so profit from lactose whenever it becomes available. This is another example of the Toyota-like "just-in-time" principle.

*Induction:* This is a word generally used to refer to the positive effect of a metabolite on the biosynthesis of a protein, as in the case of lactose, in which lactose-degrading enzymes are produced only when lactose is available in the medium.

Thus, there are two levels of feedback regulation by end products on a metabolic pathway: One regulates the flow of material through the pathway, and the other regulates the concentration of each protein (enzyme) that catalyzes the chemical reactions that comprise the pathway. Both of these levels contribute to metabolic homeostasis.

*Homeostasis:* This useful word was coined by the American physiologist Walter Cannon in 1926 to refer to the idea, first introduced in the 19th century by Claude Bernard, a French physiologist, that a stable internal chemical environment (*"milieu interieur,"* as Bernard called it) is the essential precondition for a free-living organism. This principle applies to all living things, from single-celled bacteria to complicated, multicellular plants and animals (including humans). To grow, all free-living organisms need to acquire nutrients and chemically transform them into building blocks and energy for growth and reproduction, and to also maintain control over all of this chemistry, such that everything happens in an orderly way, even when their environment is constantly changing. Inhibition and activation, repression and induction; are all means by which homeostasis is achieved.

In summary:

▶ *In regulatory biology, the words "inhibition" and "activation" are used to refer to regulation at the level of enzyme activity; "repression" and "induction" are used to refer to regulation at the level of enzyme biosynthesis.*

The feedback inhibition by a pathway's end product is fast, and usually quickly reversible. Under circumstances in which the environment is changing relatively rapidly, this mechanism is ideal for keeping the levels and flow of metabolites steady, producing the *milieu interieur* required for efficient organism function and viability. This process is also probably sufficient to keep the many different pathways that contribute to growth balanced so that no one pathway steals substrates or energy from the others.

Under circumstances in which the environment is changing on a longer time scale, the induction–repression mechanism can change the protein content of a cell on a semipermanent basis. An excellent example is the *E. coli* bacterium itself. In nature, it is found in two different environments: the intestines of animals and the soil. The former is a relatively high-temperature environment ($37°C$), in which lactose is often present and plentiful, especially when the animal host is an infant. The latter is a much more variable, low-temperature environment, in which lactose is rarely, if ever, encountered. Many of the genes that function in these two alternate environments are controlled by repression and induction together. *E. coli* appears to have evolved something close to a differentiation pathway, which enables the metabolic abilities of the bacterial cell to transition between these two environmentally adapted states, with different cellular protein compositions, as bacteria move from the animal gut to the soil and vice versa.

Not surprisingly, then, the mechanisms of induction and repression have turned out to be fundamental to the ability of organisms to deal with longer-term challenges, including the ultimate challenge of differentiating their cells into the many different types that make up a multicellular animal. The way in which proteins are made and how their synthesis is controlled is the main focus of much of the remainder of this book.

**Aaron Novick (1919–2000)** was an American physical chemist and molecular biologist. After working on the Manhattan Project during World War II, he became a molecular biologist, working with Leo Szilard to invent the chemostat, which they used to measure mutation rates and to study the molecular biology of gene regulation in bacteria.

**Leo Szilard (1898–1964)** was a Hungarian physicist and physical chemist and later became an American molecular biologist. He was the first to conceive the idea of a nuclear chain reaction based on multiplication of neutrons and obtained secret British patents on nuclear reactors and atomic bombs. He was a friend of Albert Einstein, and he wrote the famous letter to President Roosevelt that resulted in the Manhattan Project and convinced Einstein to send it. After the war, Szilard took up biology. He invented the chemostat (with Aaron Novick) and contributed many other profound ideas to biology, spanning the field from affinity chromatography to the basis of antibody diversity.

**Jacques Monod (1910–1976)** was a French microbial physiologist and molecular biologist. During World War II, he was a hero of the French Resistance at the same time as he pioneered the study of metabolic regulation in bacteria. He introduced the idea of allosteric interactions as the mechanism of feedback regulation of protein activity. With François Jacob, he worked out the mechanism of regulation of induced synthesis of the lactose-degrading enzymes in *E. coli*. Monod and Jacob discovered and named the basic elements, including repressor, operator, and promoter. Their work established the basic principles of regulation at the transcriptional level.

**François Jacob (1920–2013)** was a French geneticist and molecular biologist who pioneered the development of bacterial genetics. With Jacques Monod, he worked out the mechanism of regulation of induced synthesis of the lactose-degrading enzymes in *E. coli*. Jacob's facility with bacterial genetics was crucial to the demonstration of the *cis*-action of operator and promoter mutants. With Sydney Brenner, Jacob first proved that unstable "messenger RNA" carries information from the genes to the ribosomes.

**Daniel E. Koshland, Jr. (1920–2007)** was an American biochemist and molecular biologist who introduced the "induced fit" model for enzyme catalysis and regulation (similar to Monod's *allosterie*), developed a theory for how protein modification can confer ultrasensitivity to ligands, and showed how protein modifications account for adaptation in bacterial chemotaxis. He was a great academic leader, consolidating and revitalizing biology teaching and research at the University of California, Berkeley. For 10 years, he was Editor-in-Chief of *Science* magazine.

**Walter Cannon (1871–1945)** was an American physiologist. He made many contributions, among them the idea of imaging digestion by feeding patients heavy metals and the idea of the "fight or flight" response to threats to an animal, and he reintroduced and named the idea of homeostasis based on the ideas of Claude Bernard.

**Claude Bernard (1813–1878)** was a French physiologist and champion of the scientific method who introduced several lasting ideas. First among them was homeostasis, which he called the constancy of the *"milieu interieur."* He also introduced the idea of blind experiments as a way to limit subjectivity in the performance of scientific measurements. His book *An Introduction to the Study of Experimental Medicine* (1865) is a masterful and prescient exposition of the practical relationship between hypothesis, evidence, and scientific truth.

# CHAPTER 6

# Phage and the Beginning of Molecular Genetics

The work of pioneering molecular biologist Seymour Benzer has become the rigorous intellectual basis for what we now call molecular genetics. Because so many of the ideas and so much of the language of modern genetics derives from this work, I describe it in considerable detail in this chapter. The initial question Benzer addressed concerned the size of the functional gene in terms of its DNA structure. He reasoned that if he could measure the frequency of recombination with enough sensitivity to reach the lower limit (which he took to be a crossover between adjacent base pairs in DNA), it should then be possible to find recombinants that lie well within the stretch of DNA that encodes a functional gene. He realized that he had discovered the perfect tool for studying the fine structure of genes in a class of bacteriophage T4 mutants, ultimately named *rII* mutants.

Phage T4 was introduced in Chapter 4 in the context of a developmental pathway (Fig. 4.3). The T4 genome, like the genomes of most phage, consists of a single, linear DNA molecule. The DNA is encapsulated in a protein shell (referred to as its "head"), which is attached to a mechanism (called its "tail") that injects the DNA into its host. When one or more phage infect the host, normal cellular metabolism is hijacked and the cellular DNA is destroyed. Thereafter, all cellular activities obey the instructions of the injected T4 genome. These instructions provide all the information needed for making 100 to 150 progeny phage particles in ∼15 min with all their associated structures. The phage particles are then put together by the assembly pathway shown in Figure 4.3. To end the process, the T4 genome directs the synthesis of proteins that cause the dissolution of the cell (a process called "lysis"), allowing the progeny to be released into the environment.

The injected T4 DNA molecule encodes the entire phage genome. There are no chromosomes in the ordinary sense, just DNA in the bacterial cytoplasm. Nevertheless, the concepts and implicit experiments for understanding phage genetics work just as well as they do for understanding the genetics of pea plants, yeast, or humans. Indeed, the relative simplicity of the T4 system highlights some general principles, as there are relatively few arcane facts required to understand what must be going on in an infected cell that is following the directions of the relatively small T4 genome.

## Gene and Locus in Bacteriophage T4

Today, we know that the T4 genome consists of a DNA molecule of about 300 functional genes. Its length is about 170,000 bp (170 kb). If the functional genes were laid end-to-end (which we now know they are), the average length of a gene would be ~550 bp, enough to encode proteins with an average length of about 180 amino acid residues. Although these statistics are interesting, they do not tell us how these molecular facts connect to the functions of these genes. To understand this, we need to connect the functions of genes more directly to the molecules that execute them.

The approach Benzer used was to study intensively the structural and functional relationships among large numbers of mutants with a common phenotype, by choosing a phenotype that was easy to study quantitatively. As it will become clear later in this chapter, this work led to a detailed understanding of the relationship between the *rII* locus of T4 and the two functional genes it contains. Benzer's analysis of the *rII* locus is the basis of the modern definition of the functional gene. The ideas and language he introduced are still used today to connect DNA sequence variation to biological function in all organisms, including humans.

The same principles of genetic analysis apply to phage T4, humans, and plants, even though the human genome is 20,000 times larger (3300 Mb) and the genome of Mendel's pea plant (*Pisum sativum*) is larger still (4300 Mb) than that of T4. This became clear in the late 1940s when the American microbiologist and geneticist Alfred Hershey and his colleague Raquel Rotman discovered and studied recombination among plaque morphology traits (which I explain in more detail below).

An important feature of T4 not shared by humans or peas is that a bacterium infected by a single phage T4 genome is essentially haploid, meaning that there is just one copy of each DNA sequence. All the other organisms we have so far discussed are diploid (i.e., their cells—with the exception of the gametes—contain two homologous copies of each chromosome). This means that in diploid organisms, phenotypes (at the organismal and cellular level) are the result of the action of two potentially different versions of the same genes. This, of course, was one of the basic insights achieved by Mendel that underlies the concepts of dominance and recessiveness.

Organisms (such as bacteria and viruses) that are normally haploid offer great technical advantages for experimental genetics research. For example, in a cross between a dominantly acting allele and a recessive one, no further analysis is needed to determine the genotypes of the progeny; the phenotype of each progeny tells you its genotype. In diploid organisms, by contrast, a progeny organism that displays the dominant phenotype could be either a homozygote or a heterozygote. And working out whether it is one or the other requires a further cross to an organism that is known to be homozygous for the recessive allele, and then examining the phenotype of the progeny. I described this procedure (called a "test cross" or "backcross") in Chapter 2.

There are many additional advantages of a phage system. First among them is scale. Crosses involving many hundreds of progeny are just about feasible with some plants and can be very hard to achieve with animals, whereas phage crosses that involve the analysis of hundreds of millions of progeny are routine. Second, the time required to do crosses is a minimum of weeks with simple animals like flies, and much longer (a growing season) with plants; in contrast, Benzer and the Italian virologist Renato Dulbecco famously described their standard regime of doing three crosses every day when they worked together in the 1950s. Third, the genome of phage is small and chemically well defined as being just a single, long DNA molecule, whereas the genomes of higher organisms contain vastly larger amounts of DNA packaged into much more complicated chromosomal structures.

## Phage Mutant Phenotypes

Before discussing in more detail the molecular genetics of phage, it is useful first to understand the main phenotypes that were studied by Benzer and others working in this field.

*Plaque Morphology:* Just as the first genetic traits studied by Mendel consisted of visible and heritable differences in the appearance of pea plants, the first phage mutants were those with visibly different growth patterns. When a small number of phage (around 100) are deposited on an agar plate together with a large excess of sensitive bacteria, the bacteria produce an even "lawn" of growth punctuated by circular, clear spots. These spots mark the places where a phage has landed, infected a bacterial cell, and produced hundreds of progeny, which killed the cell before spreading to and growing in many more neighboring cells. This cycle of infection and killing ultimately produces a local clearing in the bacterial lawn called a "plaque." Phage are easily enumerated by dilution and plating on such lawns. Each plaque represents a single viable virus particle.

Early phage workers noticed that some plaques looked different than others, and upon repropagation, the variant plaque morphology proved to be heritable, just as wrinkled and yellow pea morphology proved to be heritable traits in the peas studied by Mendel. These different plaque morphologies were the traits used for the first phage crosses by Hershey and Rotman. From these crosses, it became clear that phage genomes do recombine and that recombination frequency can be used to measure distance on a phage genetic map, just as in flies and plants.

*Conditional Lethality:* In the late 1950s, two classes of phage mutants were found that grow in one circumstance but not another; hence the name "conditional lethal." One kind of mutant's growth is conditional on temperature: These mutants grow and make plaques at 30°C but not at 42°C, and are thus called temperature-sensitive (*ts*) mutants; we encountered these before in Chapter 4. Another kind of mutant's growth is conditional on the strain of the bacterial host: These mutants grow and make plaques on one bacterial strain but not another. These are called host-range mutants. Both classes of conditional-lethal mutants played crucially important roles in the development of phage genetics and molecular biology, and more will be said about them in Chapters 9 and 10.

*Dual Phenotype of r Mutations:* The *r* mutant was one of the earliest plaque morphology phage mutants to be discovered, and it has several phenotypes. It was discovered when it was grown on the standard host *E. coli* strain, B, and its plaque morphology phenotype traced to the ability of T4*r* mutants to lyse cells faster than wild-type T4 can; "r" stands for "rapid lysis." It soon became apparent that the mutations that cause the r phenotype map to three loci that can be separated by recombination (see below), called *rI*, *rII*, and *rIII*.

Benzer discovered that T4*rII* mutants also have a host-range phenotype; they are unable to grow or make plaques on another strain of *E. coli*, called K12($\lambda$) (which we will henceforth call K), on which wild-type T4 grows normally. This property was traced to the $\lambda$ prophage in K (a prophage is a latent form of a phage that can exist in a bacterium without disrupting the cell). When this prophage was removed, T4*rII* mutants were able to grow normally on the resulting strain (called S). Furthermore, the plaques made on S lawns have normal r$^+$ phenotypes. Thus, the plaques look like wild-type T4 plaques, and lysis by *rII* mutants in S is not faster than that by wild-type T4.

To summarize the r phenotypes:

- T4*rII* grows and makes r-type plaques on *E. coli* strain B.
- T4*rII* fails to grow and makes no plaques on *E. coli* strain K.
- T4*rII* grows and makes normal (r⁺-like) plaques on *E. coli* strain S.

## Complementation and Recombination Assessments in Phage

As noted above, T4 phage is effectively haploid: Its genome contains essentially one copy of each DNA sequence. A bacterium infected with one type of phage is therefore also effectively haploid. Thus, special measures are required to produce the heterozygotes required by the implicit experiments discussed above that define both the functional gene and its location in the genome. This turns out to be quite simple: A bacterium is simultaneously infected with equal numbers of two phage genomes that differ in sequence. The result is an infected cell, the phenotype of which depends on the action of two versions of each gene, just as in a standard heterozygote. In practice, coinfection is carried out with mixtures of at least three copies of each genotype per cell, so that on average each cell has roughly equal representation of both genotypes.

*Dominance and Complementation Tests:* The experimental tests for determining dominance and complementation in phage crosses employ doubly infected cells (which are equivalent to heterozygotes) and follow the logic I have previously described in this book for other organisms. Virtually all these tests employ conditional-lethal mutations, which are phenotypically assessed by their growth. Growth measured as the number of progeny phage per infected cell resulting from a nonpermissive coinfection.

*Nonpermissive Condition:* This is a condition in which a particular conditional-lethal mutation cannot provide its normal function.

*Permissive Condition:* This is a condition in which a particular conditional-lethal mutation provides normal, or nearly normal, function.

For example, consider a test for complementation between a host-range mutant, *h1*, that grows on strain P but not on strain Q of bacteria and a temperature-sensitive mutant, *ts2*, that grows at 30°C but not at 42°C in either strain P or strain Q. The coinfection for this test would be carried out in strain Q at 42°C because these are the conditions in which neither the *h1* mutant (which works only in strain P) nor the *ts2* mutant (which works only at 30°C) function.

After a single cycle of growth, the number of progeny that emerge would be measured by a plaque assay. This assay would use strain P at 30°C, conditions permissive for both parental phage, because it is the progeny's growth, regardless of their genotype, that is of interest. A near-normal ($\sim$100) number of phage per cell supports the conclusion that the two mutations lie in different functional genes. A progeny number near the background (usually 0–5 phage per cell) would indicate that the two mutations lie in the same functional gene.

> ▸ *In tests of phage function, coinfection is done under fully* nonpermissive *conditions, whereas measures of growth are done under fully* permissive *conditions.*

***Recombination Frequency:*** Recombination in phage is defined exactly as it is in other organisms: the fraction of progeny from a doubly infected cell that have the recombinant (as opposed to parental) arrangement of alleles. Virus particles are the equivalent of haploid gametes, being the haploid derivatives of the equivalent of a doubly heterozygous diploid cell.

In a cross between two phage with different, independently scorable plaque morphology phenotypes ($h$ and $r$, as in the illustration shown in Fig. 6.1), all the possible progeny genotypes (the parents, $h\ r^+$ and $h^+\ r$, as well as the

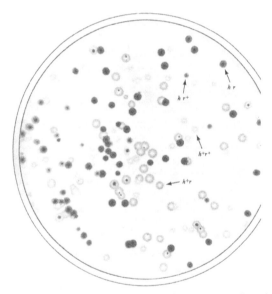

**Figure 6.1.** The results of a cross between T4 bacteriophage mutants $h$ and $r$. Plaques are made by the progeny of T4 phage coinfected with two plaque morphology mutations: $h$ and $r$. All the possible progeny genotypes (the parents, $h\ r^+$ and $h^+\ r$, as well as the recombinants, $h\ r$ and $h^+$ $r^+$) make plaques that are distinguishable on the same agar plate. (Reproduced, with permission, from Stent GS, Calendar R. 1978. *Molecular genetics: An introductory narrative*, 2nd ed. W.H. Freeman, San Francisco.)

recombinants, $h\ r$ and $h^+\ r^+$) are distinguishable on the same agar plate. Counting these four phenotypes gives the same kind of results as those obtained from recombination analyses carried out in other organisms. This means that the frequency of recombination can be determined directly using the standard definition

$$F_{rec} = (\text{number of recombinant gametes})/(\text{total gametes}).$$

However, plaque morphology crosses are limiting because most of the interesting traits in phage are conditional lethals of one sort or another. These all have similar mutant phenotypes: failure to grow under some nonpermissive condition, as described above.

**Selective Crosses:** Recombination between two conditional-lethal mutations produces two kinds of recombinants: a double mutant and a phage that carries neither mutation, which will have the wild-type phenotype, allowing it to grow in circumstances that are nonpermissive for either of the parental conditional lethals. If one makes the rather mild assumption that the recombination mechanism will produce the double-mutant and the wild-type recombinants with equal frequency, doubling the number of wild-type recombinants provides a good approximation of the total recombinant frequency. This assumption had already been shown to hold where it has been tested—namely, in crosses among distinguishable plaque morphology mutants. Now the problem of counting recombinants can be solved simply. Two measurements are required: the number of progeny, as assayed under totally permissive conditions (total phage); and the number of recombinants, as assayed under totally nonpermissive conditions (conditions under which one-half of the recombinants can grow).

To illustrate this idea, consider a cross between the two mutations $(h1 \times ts2)$ that I used as examples above. To measure recombination frequency between these two strains, they would be coinfected in strain P at $30°C$ (i.e., using fully permissive conditions) to allow any phage, regardless of genotype, to grow, making this a test of recombination and not function. After a single cycle of growth, their progeny are collected and their total number determined by diluting and plating them under fully permissive conditions (strain P at $30°C$). This assay yields the denominator of $F_{rec}$. To assess the number of recombinants, plating is then done under conditions (strain Q at $42°C$) in which only wild-type recombinants $(h1^+\ ts2^+)$ can grow. If we assume that a similar number of $h1\ ts2$ and $h1^+$ $ts2^+$ recombinants are generated in the cross, then multiplying the number of wild-type recombinants by 2 yields the numerator of $F_{rec}$.

So, for a selective phage cross,

$$F_{\text{rec}} = 2 \times (\text{number of wild-type phage})/(\text{total progeny phage}).$$

▶ *Tests of phage recombination require coinfection under fully permissive conditions, and then finding the number of wild-type recombinants produced in one cycle of growth by plating progeny on nonpermissive conditions and the total number of progeny by plating under fully permissive conditions.*

Making such selective crosses enormously expands the sensitivity with which we can measure recombination. In a nonselective cross, one has to count all the progeny, so recombinant frequencies that are less than one in a few hundred become burdensome to identify. Studies at this level of resolution also become impractical because the numbers of recombinants obtained in a few hundred progeny do not come close to the sensitivity that would be required to resolve the fine structure of genes at the resolution of individual base pairs, the resolution that Benzer achieved.

But with phage, all one needs to do is to carry out the crosses in a tube containing 1 ml of solution containing $10^8$ bacteria, which will produce $10^{10}$ progeny after coinfection. The progeny are then diluted appropriately for the permissive plates, and diluted two or three orders of magnitude less for the nonpermissive plates, making it easy to detect recombination down to the mutational background, down to ~1 in $10^7$ for mutations with low reversion rates. It was this feature that made Benzer's studies possible and brought some measure of reconciliation between the idea of a locus and that of the functional gene.

## Genetic Fine Structure

What can be learned from "running the map into the ground," in Benzer's memorable phrase? First of all, we can hope to find out whether all the mutations that map at a locus (i.e., at a stretch of DNA) belong to the same functional gene, because we can do two tests with a pair of mutants: We can measure recombination, which tells us how close together they are in the genome; and we can measure complementation, which tells us whether they specify the same or different functions. Today, we know that protein and RNA molecules execute most cellular functions and that each is encoded by a DNA sequence that comprises a functional gene. In the 1950s, this was still a supposition and not yet a fact.

The first thing needed to carry out a fine-structure-mapping program is a very large number of mutations. Benzer ultimately analyzed more than 2400 *rII* mutations. He isolated many more that he elected not to use. In the *rII* system, as in most other systems, not all mutations are equally easy to study. Like Mendel, Benzer was faced with choices concerning which mutations to study. Mendel chose traits to study on the basis of their breeding true over many generations. Benzer chose mutations to study on the basis of how well they behaved in his system.

Three issues were of particular importance:

***Strength of Phenotype:*** Mutations vary a lot in the strength of their phenotype. In our discussion of types of mutant phenotypes in Chapter 2, null mutations were introduced as ones that fail completely to provide a function. Historically, many geneticists noticed a great deal of variation in the strength of phenotypes caused by mutations in the same functional gene. Many, including Benzer, referred to this as "leakiness" of the phenotype. Today, extreme variation in leakiness is easy to rationalize: The molecular consequences that mutations have for the encoded protein are different for different kinds of mutations, a subject I will discuss extensively in the later chapters of this book. Leakiness is a big problem for complementation tests because partial function introduces ambiguity. Thus, Benzer measured leakiness for all of his phage mutants. He used the frequency with which a viable phage will emerge from a single infection of an *rII* mutant in strain K, the nonpermissive host; he called this the "transmission coefficient." He then excluded leaky mutants from further analysis. Even today, when embarking on a mutational analysis, most geneticists will avoid leaky mutants if they possibly can.

***Revertant Frequency:*** Mutations vary a lot in the rate at which they return (revert) to the normal genotype. A lower rate of reversion means higher resolution in recombination-mapping experiments because the number of plaques on K, the nonpermissive host, is the critical parameter. Stocks of some *rII* mutants contain revertants that make plaques on K at frequencies as high as $1/10^4$; these would produce a background in crosses that would severely limit the ability to detect recombination frequencies to those well above $1/10^4$. Most *rII* mutants revert at much lower frequencies, between $1/10^6$ and $1/10^8$, and some fail to revert at all. This behavior is typical of large collections of mutations in many different genetic systems. Benzer, like Mendel, used the ones that he could work with.

***Deletion Mutations:*** Extensive deletion mutations do not revert at all. In the discussion of null mutations in Chapter 2, I referred to the usefulness of experimentally deleting entire functional genes. However, deletion mutations also occur naturally. Mutations that are not deletions and that fail to recombine with a deletion logically must lie within the extent of the deletion. Benzer found many deletions among his nonreverting *rII* mutations.

In a cross between a nonreverting deletion and a reverting mutation, revertants of the reverting mutation provide the only background that limits resolution. The potential effects of background reversion rates turned out not to be a problem for the $rII$ system. Benzer found the lower limit for recombination to be about $2 \times 10^{-5}$, well above the reversion frequencies of most $rII$ mutations he used. This minimum recombination distance Benzer named the "recon." Only values of $F_{rec}$ at or above this value support the conclusion that recombination had occurred. There was little ambiguity when reaching this conclusion because the mutations Benzer studied had reversion frequencies about 20-fold lower than the recon value. However logical it may have been, the term "recon" never really caught on.

▸ *The genetic definition of a deletion is a nonreverting mutation that fails to recombine with two other mutations that recombine with each other.*

With a subset of well-characterized $rII$ mutants, it became possible for Benzer to map his chosen mutations without carrying out nearly 6 million ($2400 \times 2400$) pairwise crosses. He first crossed the mutants with a set of deletion mutations, each with different end points that span the entire $rII$ locus. Pairwise crosses were needed only for mutations that were already known to be in the same or neighboring deletion intervals. This is the way in which very many fine-structure maps in many different organisms were made before DNA sequencing became cheap and easy.

## Locus and Functional Gene Reconciled

By any standard, the fine-structure map of T4's $rII$ locus was a tour de force and a thing of beauty.[1] It was extremely accurate, as subsequently proven by DNA sequencing 20 years later. Many of the $rII$ locus mutations that are most closely linked (meaning that recombination between them is at or near the minimum of $2 \times 10^{-5}$) are indeed, as anticipated by Benzer, alterations in adjacent base pairs. This result not only justified the thinking behind Benzer's recon; it also justified the assumption that the minimal mutable unit is indeed a base pair. Benzer called this unit a "muton," another word that never caught on.

Science, however, does not reward or remember beautiful work per se. Benzer's work is not remembered and revered today by specialists because his genetic map turned out to be correct down to the smallest detail. Instead, it is remembered by the larger scientific community because it reconciled the ambiguity

---

[1] Benzer S. 1962. The fine structure of the gene. *Sci Am* **206**: 70–80.

and confusion surrounding two concepts: the genetic locus (as defined by recombination then, and by DNA sequencing now) and the functional gene (defined by complementation, then and now).

This wonderful result emerged when the many mapped *rII* mutations were tested for whether they complement each other in a pairwise fashion. Although the distribution of sites at which *rII* mutants occur across the entire locus is pretty uniform, the mutations fell into two complementation groups, which Benzer named *rIIA* and *rIIB*, each of which comprised about half of the total mutations. Remarkably, the *rIIA* mutations were all to the left of a genetic position (called the boundary) and the *rIIB* mutations were all to the right of it. The recombination distance between the rightmost *rIIA* mutation and the leftmost *rIIB* mutation was very small, much smaller than the distance between either of these mutations and the most distant mutation in the same complementation group. Subsequent DNA sequencing showed that only 9 bp separate the rightmost *rIIA* mutation and the leftmost *rIIB* mutation, as compared to 1449 bp (the total length of *rIIA*) and 939 bp (the total length of *rIIB*).

The relationship between a sequence of DNA at a locus and a functional gene at that same locus was clarified by this result. Benzer showed that the DNA sequence of the *rII* region is continuous and mutable along the entire length of this locus, and that mutations within this locus produce similar phenotypes. However, his complementation analysis showed that two independent functions are encoded by the subregions that fall to either side of the aforementioned boundary, each of which is somehow required to produce the r$^+$ function. On the assumption that the DNA sequences at the *rII* locus encode protein molecules, as we now know they do, a simple model would be that *rIIA* and *rIIB* each encode a separate molecule, both required to carry out the r$^+$ function. If this were the case, and if the DNA sequence were indeed continuous across the locus and boundary, then the position of the beginning and the end of each coding segment must also be encoded in this sequence, as indeed they are.

## The *cis–trans* Test and the Cistron: A Formal Definition of the Functional Gene

Benzer produced a formal definition of the functional gene based on an implicit experiment that had been independently proposed a few years earlier by the American geneticist Edward Lewis and the Italian-born geneticist Guido Pontecorvo. This implicit experiment (called the *cis–trans* test) is important because it completely generalizes the definition of the functional gene and makes it independent of its molecular structure.

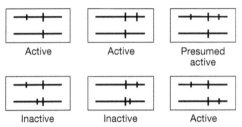

**Figure 6.2.** Phenotype and the configuration of mutations. This diagram from Seymour Benzer's study illustrates the difference between the expected phenotypes of a double heterozygote, depending on whether the mutations are in a *cis* or *trans* configuration. The two horizontal lines represent the two parental sequences at the locus; the dotted vertical line, a boundary between functional genes; and the solid vertical lines, the positions of mutations in various arrangements. Active/inactive refers to gene function. (Reproduced from Benzer S. 1955. *Proc Natl Acad Sci 41:* 344–354.)

The implicit experiment is illustrated in a simple diagram (see Fig. 6.2), taken from Benzer's 1955 paper.[2] It shows several possible positions of two mutations in a diploid heterozygote being tested for its phenotype. In this illustration, the two horizontal lines are the two parental sequences, or DNA strands, at the locus; the dotted vertical line represents a boundary between functional genes; and the solid vertical lines indicate the positions of mutations, in various arrangements. The important comparison is that between the two rightmost panels (top and bottom) in Figure 6.2. These panels show one copy each of two different mutations: In one arrangement (top), the mutations are *cis* to each other (i.e., they are on the same DNA strand from the same parent); in the other (bottom), they are *trans* to each other (i.e., on opposite DNA strands, one from each parent). The only difference between them is their configuration; in terms of composition (presence and absence of mutations), they are identical. When we compare the phenotypes of these two double heterozygotes, it is called the *cis–trans* test. It is a formalization and a molecular interpretation of the complementation test.

Benzer named the unit of function—the stretch of DNA within which all recessive mutations fail to complement—the "cistron." If the results of the *cis–trans* test are as shown in Figure 6.2, the two mutations are in different cistrons; if in *trans* mutations fail to provide function, they belong to the same cistron. Unlike the recon and muton, the cistron neologism has stuck. Even though the formal definition of this word does not depend on knowing about the detailed structure of a gene and the information it contains, the cistron has come to

---

[2] Benzer S. 1955. Fine structure of a genetic region in bacteriophage. *Proc Natl Acad Sci* **41:** 344–354.

represent the DNA sequence that encodes a single polypeptide chain, together with all the information required for its translation. Readers will find it defined in this simple way in most textbooks and dictionaries, but most geneticists understand the value of the formal definition provided by Benzer because it does not depend on molecular detail.

The reader will have realized that constructing the *trans* double heterozygote is simple: It is the hybrid produced by a cross between the individual mutations to be tested. The *cis* double mutant is more challenging to isolate, as it requires a double mutant to be recovered in which a recombinant event has taken place between the two mutations, one in each cistron, which can then be crossed with wild type. This is very difficult to arrange when the recombination frequency between the two mutations is one in many thousands. Although, in principle, assessing the phenotype of the *cis* arrangement should always be performed, even Benzer assumed that this arrangement would always provide function from the unmutated strand, given the recessive nature of the individual mutations concerned (see the upper left and middle panels of Fig. 6.2).

For dominant regulatory mutations, testing the *cis* configuration turned out to be extremely important. The return in information more than justified the substantial efforts required to construct the relevant double mutants. I will discuss this in considerable detail in the context of regulation in Chapter 8.

The features of a functional gene (cistron) that can be inferred from Benzer's studies of *rII* can be easily generalized to the entire genome of T4, and that of any organism: Functional genes are stretches of DNA sequence that encode other molecules. Today, we know that cistrons encode RNA molecules, some of which remain as RNA molecules but most of which are used to create polypeptide chains.

## Many Cistrons Can Contribute to a Single Biological Function

It is important to emphasize that several different cistrons might be involved in the production of a single functional protein structure. Enzymes often consist of several different polypeptides that together perform the same cellular function. As one might expect, mutations that result in the loss or aberrant function of any one of several constituent polypeptides result in a similar phenotype, because they each cause the loss of the same enzyme's function.

For example, the eukaryotic RNA polymerases that read DNA into messenger RNA (mRNA) (including those of humans and yeast) consist of a minimum of 12 different subunits. Each polypeptide is produced by a different cistron. Therefore, if a heteroallelic diploid in yeast contains mutations in two different cistrons of

RNA polymerase, all is well. However, if a heteroallelic diploid contains two mutations that each lie in the same cistron, then all is not well, and the diploid will have the mutant phenotype—the inability to read DNA to make mRNA.

Thus, we are left with the final conclusion to Benzer's definition of the functional gene. The functional machinery of cells consists of very many interchangeable parts (protein and RNA molecules), some of which work together in close association with one another. All of this molecular machinery is encoded in the DNA sequence of an organism's genome. The cistron (defined by complementation tests) refers to the lowest level of interchangeable parts, which, in proteins, is a polypeptide chain.

## INTRODUCTORY BIOGRAPHIES

**Seymour Benzer (1921–2007)** was a solid-state physicist by education. He took up the study of fine-structure genetics in bacteriophage T4 shortly after finishing his PhD in physics; this is the work that led to modern understanding of the functional gene. After 20 years, Benzer took up the study of behavior in *Drosophila*. He found Mendelian genes that affect sexual behavior, memory, circadian rhythm, and longevity.

**Alfred Day Hershey (1908–1997)** was an American bacteriologist and geneticist. He was a founder of the nascent field of molecular biology, having taken up the study of bacteriophage with Salvador Luria and Max Delbrück, with whom he won the Nobel Prize. His lasting accomplishments were an elegant experiment that showed that the genetic material of T4 is DNA, not protein; the discovery of recombination in T4 and construction of the first phage genetic maps; and the discovery of the cohesive single-stranded ends of phage λ. A man of few words, he exerted intellectual leadership in private conversation and through his excellence as an editor.

**Raquel Rotman (Sussman) (b. 1921)** is an American geneticist who worked with Alfred Hershey and discovered recombination in T4 and made the first genetic maps of T4. Later, under her married name (Sussman), she published the discovery, with François Jacob, of a temperature-sensitive mutant of the λ repressor that allowed λ to lysogenize normally at 30°C, making lysogenic bacteria that could be induced by shifting the temperature to 42°C. This result was of critical importance in understanding negative control of phage λ.

**Edward B. Lewis (1918–2004)** was an American geneticist who founded the field of developmental genetics in his analysis of the bithorax loci, the founding member of a conserved gene family (the homeotic genes) that controls the body plan of all animals, from insects to mammals. In his extensive genetic studies of developmental mutants, he addressed the issues around multiple alleles at complex loci with the result that he formulated the original version of the *cis–trans* test, together with Guido Pontecorvo.

**Guido Pontecorvo (1907–1999)** was an Italian-born Scottish geneticist whose career began with a PhD with Hermann Muller, who introduced him to the issues around the nature of genes. He was one of the first of the classical fly geneticists to recognize the experimental virtues of microbial systems for studying genetics. He developed the ascomycete fungus *Aspergillus nidulans*, and began to study fine-structure genetics and to address the problem of multiple alleles at a locus. In this he was outdone by Benzer, largely because of the practical advantages of the T4 system. He was a major intellectual leader, building the University of Glasgow into a major center of genetic research.

CHAPTER 7

# Transcription, Translation, and the Genetic Code

The analysis that led to the definition of the cistron also led to the realization that functional genes do not just consist of the sequence of DNA nucleotides that are read by the protein synthesis machinery; this sequence also encodes signals that indicate the beginning and end of the polypeptide chain. Two quite different lines of evidence led to the elucidation of the genetic code and how living cells decode it. One line was fundamentally biochemical in nature, and it produced the table of correspondences between the amino acids and the three-base "codons" that encode them in the DNA sequence (see Fig. 7.3 below). This work, elegant and important as it was, did not involve genetic analysis, and it is not described further here.

The second line of evidence, which mainly concerns how cells read the code rather than the code itself, laid the foundation for modern molecular genetics. I describe it here in some detail, because, as with the work of Mendel and Benzer, the ideas, concepts, and vocabulary involved are central to the interpretation of DNA sequence today.

Molecular biologists introduced a set of important words to refer precisely to the steps of the decoding process. These steps, which were only surmised in the 1960s, have been studied intensively by molecular biologists ever since and are now very well understood. However, the detail of how we came to understand the decoding process is not required to understand the results of genetic analysis or to interpret DNA sequence. For this reason, I have confined the following outline to only a few essential concepts and words.

I assume most readers will have encountered the basics of molecular biology before; they have become central to the teaching of biology in schools and colleges. Surprisingly, there remains some diversity in how some very fundamental issues are described. For example, there is still confusion in the way that the "coding strand" is defined. My goal here is to introduce as simply and as rigorously as I can the vocabulary and conventions that I (and most geneticists I know) use today.

Readers who need more detail may wish to consult any standard genetics, molecular biology, biochemistry, or cell biology textbook, keeping in mind that diversity in some definitions still remains. I suspect many readers may find that these books provide substantially more detail and specialized vocabulary than is necessary or even useful for understanding what follows in this book.

## The Flow of Information from DNA to Protein

Biological information is stored in DNA as the sequence of bases. The famous double helix of DNA (Fig. 7.1) is held together by base-pairing, such that if one strand has an A base (adenine), then the other has a T (thymine); if one has a G base (guanine), then the other has a C (cytosine). The DNA molecule is symmetrical around the helical axis, and the strands run in opposite directions, for which the chemical notation 5′ and 3′ remains in general use. From the sequence point of view, then, a short DNA sequence might look like this:

$$5' \ldots \text{AGCTTGGCATTACTG} \ldots 3'$$
$$3' \ldots \text{TCGAACCGTAATGAC} \ldots 5'$$

The information in both strands is the same, although the sequence is chemically different. The base-pairing rule allows DNA molecules to be replicated, producing two identical copies. However, decoding the information contained in DNA involves an intermediate step of copying one strand of a very long segment of DNA into a very similar molecule called RNA. RNA, like DNA, is a nucleic acid that consists of a sugar-phosphate backbone (almost identical to the one in DNA) to which bases are attached. Three of the bases (A, G, and C) are the same as those found in DNA; the remaining base is a minor variation of thymine called uracil (U). U, like T, pairs exclusively with A. An RNA sequence that is complementary to a DNA sequence will pair with it to make a hybrid DNA–RNA double helix. An RNA copy of the lower strand of the DNA sequence shown above would look like this:

$$5' \ldots \text{AGCUUGGCAUUACAC} \ldots 3'$$

Clearly, this is chemically a different molecule; it is single-stranded, contains U instead of T, and has a slightly different backbone—yet it contains the same information as the upper strand of the double-stranded DNA molecule. There are two crucial elements to this decoding process. When making RNA from DNA, a starting point has to be chosen to set where the copying begins, and the symmetry of

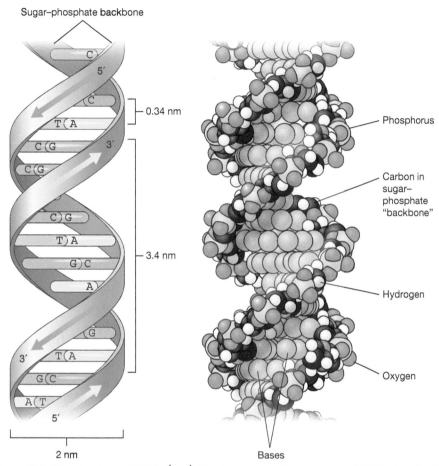

**Figure 7.1.** The structure of DNA. *(Left)* The double-stranded structure of DNA, showing the chemical structure of the sugar-phosphate backbone and the base-pairing of nucleotides. *(Right)* A schematic showing in more detail the base-pairing of nucleotides and the opposite orientation of the two complementary strands.

the double strand is broken because only one of the two strands has the coding sequence in the form that the translational machinery can read.

The decoding of DNA also occurs in two phases, called "transcription" and "translation" (see Fig. 7.2). Keeping these separate in one's mind when thinking about DNA sequence is important. A simple way to remember the difference is to recall that the English word "transcription" means "to copy," usually without changing language or meaning. By contrast, the English word "translation" refers to the conversion of text or speech from one language into another.

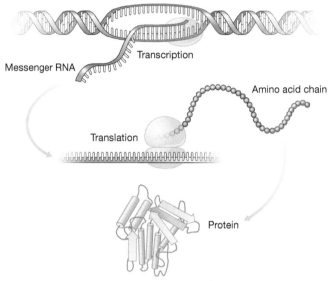

**Figure 7.2.** Overview of transcription and translation. (*Top*) A double-stranded DNA molecule is unwound to allow transcription: the copying of its nucleotide sequence information from one DNA strand into mRNA. The strand of mRNA is then decoded in a process called translation, which converts the mRNA sequence into the amino acid sequence of a protein. The translation structure shown is called a ribosome.

*Transcription:* This refers to the process by which cells decode DNA by making an RNA copy of the base sequence of one of the complementary DNA strands, as illustrated in Figure 7.1. The cellular machinery involved in transcription is complex. It includes proteins, called RNA polymerases, that perform the transcription. It also contains other factors that can read various signals contained within the DNA. These signals indicate where transcription should start, in which direction it should proceed (i.e., which DNA strand should be copied), and where transcription should stop. I discuss these signals in more detail later in this book.

*Messenger RNA:* Messenger RNA, commonly called mRNA, is the final molecule created from the process of copying one of the two DNA strands that contains the coding sequence itself. mRNA is the template used to create a polypeptide chain by the translation machinery.

In eukaryotes, the initial RNA product generated by transcription (the primary transcript) includes noncoding "intervening sequences" that interrupt the coding information. These sequences need to be removed before the mature mRNA is presented to the decoding translation apparatus. The process that removes these sequences is called "splicing," and in 1978, Walter Gilbert introduced the words "exons" and "introns" to denote the coding sequences that are retained and the intervening sequences that are removed, respectively.

*Exon:* This word refers to any segments of a gene that are retained during RNA splicing and presented to ribosomes for translation.

*Intron:* This word refers to the segments of sequence that are removed during RNA splicing. These segments can be very long: A primary transcript can sometimes be many thousands of bases in length because of the presence of numerous long introns.

In bacteria, the primary transcript created from DNA often directly functions as the mRNA. A crucial feature of mRNA biology is that mRNA molecules are unstable. With some notable exceptions, mRNA molecules last for only a fraction of a cell generation, so that continuing transcription is required for the continuing production of most proteins.

*Translation:* As mentioned briefly already, translation refers to the actual decoding of the base sequence of an mRNA molecule into the polypeptide chain encoded by each cistron. This process is governed by the genetic code. A cell's translational machinery (consisting mostly of a complex cellular structure called the "ribosome") is even more complex than its transcriptional machinery. The ribosomes recognize newly synthesized mRNAs and scan their sequence looking for the signal to start decoding. This signal (with rare exceptions) is the first occurrence of a three-base sequence, ATG, which is translated into the amino acid methionine (abbreviated Met or M; see Fig. 7.3 for the code for each amino acid). The 64 possible three-base sequences for each amino acid are called codons. Each amino acid in the chain is determined by the identity of the next codon, with no spaces. When the ribosome encounters any of the three codons TAA, TAG, or TGA (called stop codons), translation terminates, and the polypeptide chain is released from the ribosome.

*Degeneracy:* A property of the genetic code is that most amino acids can be encoded by more than one codon (see Fig. 7.3). Specifically, arginine (Arg), serine (Ser), and leucine (Leu) can be encoded six different ways; valine (Val), proline (Pro), threonine (Thr), alanine (Ala), and glycine (Gly), four different ways; isoleucine (Ile), three different ways; and tyrosine (Tyr), histidine (His), glutamine (Gln), asparagine (Asn), lysine (Lys), aspartic acid (Asp), glutamic acid (Glu), and cysteine (Cys), two different ways. The code is thus "degenerate," the mathematical term for such a many-to-one correspondence. Only tryptophan (Trp) and Met are encoded uniquely by a single triplet codon, the latter by the initiator codon ATG. This means that translation is necessarily a one-way process. The amino

| | T | C | A | G |
|---|---|---|---|---|
| **T** | TTT Phe **F**<br>TTC Phe **F**<br>TTA Leu **L**<br>TTG Leu **L** | TCT Ser **S**<br>TCC Ser **S**<br>TCA Ser **S**<br>TCG Ser **S** | TAT Tyr **Y**<br>TAC Tyr **Y**<br>TAA *stop* —<br>TAG *stop* — | TGT Cys **C**<br>TGC Cys **C**<br>TGA *stop* —<br>TGG Trp **W** |
| **C** | CTT Leu **L**<br>CTC Leu **L**<br>CTA Leu **L**<br>CTG Leu **L** | CCT Pro **P**<br>CCC Pro **P**<br>CCA Pro **P**<br>CCG Pro **P** | CAT His **H**<br>CAC His **H**<br>CAA Gln **Q**<br>CAG Gln **Q** | CGT Arg **R**<br>CGC Arg **R**<br>CGA Arg **R**<br>CGG Arg **R** |
| **A** | ATT Ile **I**<br>ATC Ile **I**<br>ATA Ile **I**<br>ATG Met **M** | ACT Thr **T**<br>ACC Thr **T**<br>ACA Thr **T**<br>ACG Thr **T** | AAT Asn **N**<br>AAC Asn **N**<br>AAA Lys **K**<br>AAG Lys **K** | AGT Ser **S**<br>AGC Ser **S**<br>AGA Arg **R**<br>AGG Arg **R** |
| **G** | GTT Val **V**<br>GTC Val **V**<br>GTA Val **V**<br>GTG Val **V** | GCT Ala **A**<br>GCC Ala **A**<br>GCA Ala **A**<br>GCG Ala **A** | GAT Asp **D**<br>GAC Asp **D**<br>GAA Glu **E**<br>GAG Glu **E** | GGT Gly **G**<br>GGC Gly **G**<br>GGA Gly **G**<br>GGG Gly **G** |

**Figure 7.3.** The genetic code for amino acids. The amino acid code summarized in a table that shows both the three-letter and one-letter abbreviations for each amino acid. Stop codons are flagged, and the single start codon is ATG (Met, M).

acid sequence is uniquely determined by a given nucleotide sequence, but very many different nucleotide sequences could encode a given amino acid sequence.

*Transfer RNA:* Another component of the translational machinery is a small RNA molecule called a transfer RNA (tRNA). These are the pieces of the translation machinery that are directly responsible for matching the correct amino acid to its codon. Francis Crick, the British biophysicist and codiscoverer of the helical structure of DNA, in a tour de force of theory, predicted the properties of these molecules before their molecular nature was established. At one end of these molecules (which are sometimes still called "adaptors," as Crick named them), an amino acid is covalently attached; at the other end, they display a three-base sequence, called the "anticodon," that uniquely recognizes codons using the same Watson–Crick base-pairing rules that hold together the DNA double helix.

*Coding Strand:* Today, this refers to the DNA strand that has the same sequence as an mRNA after transcription. Unfortunately, there is a good deal of nomenclature confusion in the literature concerning this word. Until quite recently, some textbooks had incompatible definitions, confusing the template strand with the coding strand, as it is defined today.

One way in which to avoid confusion is to keep in mind that everything works according to the Watson–Crick rules of complementary base-pairing. Thus, if the coding strand has the same sequence as the mRNA, then the transcription machinery must have copied the mRNA from the other strand, which is now generally referred to as the "template strand." DNA is symmetrical around its helical axis, so the strands run in opposite chemical directions. This is indicated by molecular biologists using the chemical notation for the DNA chains: The ribosomes scan the mRNA in the $5'$ to $3'$ direction, so the convention is to write the mRNA sequence with the $5'$ end on the left. This way, reading toward the right, we see the codons on the coding strand in the same order as ribosomes do. This means that the template strand of the DNA, from which the mRNA is actually copied, has to be written $3'$ to $5'$. An example of the directionality of each DNA strand is provided below.

DNA coding   $5'$...ATG CAT CAT CAT CAT  CAT GAT GAT TAG...$3'$
DNA template   $3'$...TAC GTA GTA GTA GTA GTA CTA  CTA ATC...$5'$

Transcription (from template strand)

mRNA   $5'$... AUG CAU CAU CAU CAU CAU GAU GAU UAG...$3'$
Peptide        Met His His His His  His Asp Asp Stop

In the example above, only a tiny fragment of a much longer DNA and mRNA molecule is shown. Recall that mRNA replaces thymine with the chemical variant called uracil, as do all RNAs. Readers will also frequently find the codon table (which, after all, is about RNA, not DNA) written with U instead of T. These are details one needs to be aware of, but I prefer the modern habit of using T for transcribed nucleic acid sequences, unless there really is a reason to be chemically fastidious.

**Reading Frame:** Because the code is read continuously, three bases at a time, a given sequence can have three alternative interpretations, called "reading frames." The sequence of the mRNA in the example given above, if we remove the ATG, is $5'$ ...CATCATCATCATCATGATGATTAG... $3'$. The different reading frames that can be obtained from this sequence are

1. CAT CAT CAT CAT CAT GAT GAT **TAG**
2. C ATC ATC ATC ATC ATG ATG ATT AG
3. CA TCA TCA TCA TCA **TGA TGA** TTA G

Clearly, these are all the same sequence: I have only spaced them differently to indicate the three possible reading frames (with stop codons highlighted in bold). To initiate translation, ribosomes scan for the ATG (usually accompanied by a specific ribosome-binding sequence) and, having found it, continue reading triplets in order from there on. Therefore, in most situations, the position of the first ATG sets the reading frame.

*Open Reading Frame:* Open reading frames (ORFs) are reading frames that are not interrupted by stop codons, and which can therefore continuously encode a polypeptide of the appropriate length. Note, in the example above, that even though reading frame 2 is open, the other reading frames have stop codons in them. Proteins typically contain 100 amino acids or more: The average size of a human protein is about 400 amino acid residues, and the largest known (called titin) is 34,350 amino acid residues long.

An ORF is therefore typically 100 codons (300 bases) or more. Given that there are 64 possible codons, three of which are stop codons, the probability of an ORF of 100 codons occurring by chance is quite small. At each step of synthesis, the probability of *not* finding a stop codon in a truly random sequence is $1 - 3/64$, or 0.953. The probability of avoiding stop codons by chance 100 times in a row is thus

$$(1 - 3/64)^{100} = 0.953^{100} = 0.008.$$

For an average protein (400 codons), the probability is very much smaller:

$$(1 - 3/64)^{400} = 0.953^{400} = 4.3 \times 10^{-8}.$$

As a result, one of the simplest and most powerful tools in the analysis of DNA sequence is to search for segments that contain lengthy ORFs, the presence of which strongly indicates that a polypeptide is encoded there.

## Molecular Taxonomy of Simple Mutations

In previous chapters, I discussed genetic variation and its role in evolution and disease. With just a rudimentary outline of how DNA stores information and how it is decoded to specify the proteins that do most of the work of cells, we can understand this genetic variation in molecular terms. What follows is a simple classification of the many kinds of mutations and variations in sequence that are possible, and some of the genetic properties of each.

*Point Mutations:* Mutations that define a single point on the recombination-based genetic map of an organism's genome are called "point mutations." Point mutations can recombine with all other point mutations and generally involve a change in only a single base pair in the genomic DNA sequence; this definition includes deletions or insertions of one base pair. Occasionally, changes (including insertions or deletions) to two adjacent base pairs can occur that have all the genetic properties of a point mutation. A very important genetic property of point mutations is that they can revert to the wild-type sequence, through the reversal of the base-pair substitution, addition, or deletion that produced them.

Point mutations can be distinguished genetically from deletion mutations in organisms in which finely detailed genetic mapping is possible (such as in the bacteriophage T4, discussed in Chapter 6). In principle, all point mutations can recombine with each other; in his studies of T4, Seymour Benzer found many instances in which recombination had occurred between what turned out to be adjacent nucleotides. Other researchers had observed "multipoint mutations" (i.e., mutations that failed to recombine with mutations that recombine with each other). These mainly turned out to be deletion mutations, as also found in the T4 *rII* system (see Chapter 6). Extensive deletions (those of more than a few base pairs) cannot revert to the wild-type sequence. However, reversal of the phenotypes of deletions is possible (although rare) through a process called "suppression," which I discuss next in Chapter 8.

When point mutations are discussed in the context of their function, rather than of their position, they are usually referred to according to their functional subclasses. These subclasses include the following.

**Substitution Mutations:** These mutations comprise the simplest class of point mutations. The term is restricted to describing the change of one base to another. There are useful functional distinctions among substitution mutations, which include the following.

**Missense Mutations:** These are base changes that result in a change of one amino acid to another in the encoded polypeptide chain. As already mentioned above, there are 64 possible codons, three of which cause chain termination, leaving 61 to encode 20 different amino acids.

**Synonymous Mutations:** These base changes cause a change in codon, but not in the encoded amino acid. With the exception of methionine (ATG) and tryptophan (TGG), all the amino acids can be encoded by more than one codon. Some (arginine, serine, and leucine) are encoded by as many as six different codons. An example would be a mutation that changes the codon CGA (which encodes arginine) to AGA (which also encodes arginine). Thus, the DNA sequence is changed, but the encoded polypeptide is not. Synonymous changes have become extremely important in molecular evolution studies, because they allow us to estimate the frequency of base-pair mutations that are expected to have little or no consequence (the amino acid is not changed), and therefore should not be either deleterious or advantageous. This estimate provides an approximation of the mutation rate over evolutionary time in the absence of selection. Knowing this rate allows researchers to assess other variations with respect to selective pressure over evolutionary time.

The phrase "silent mutation" is sometimes used to refer to mutations that are predicted to not change the function of a protein (of which synonymous mutations are a subset) because the two amino acids involved are believed to be chemically similar. I think this usage should be avoided because such predictions are unreliable; major phenotypes have been traced to such "silent changes." In contrast, although it is conceivable that synonymous changes might produce a phenotype (e.g., by making an mRNA unstable), there are very few known instances of this. It is for this reason that synonymous mutations are still generally used to estimate the frequency of "neutral" changes in protein-coding sequences (see Chapter 13).

***Nonsense Mutations:*** These are base changes that result in a codon being changed from one that encodes an amino acid to one that encodes a stop codon (TAA, TAG, or TGA), causing translation to stop. They are sometimes also referred to as "chain-termination mutations." These mutations, especially when they occur early in the coding sequence of a cistron, are usually null mutations because they result in a complete loss of function. We will return to them later, as they have played a major role in the development of molecular genetics. In recent years, the neologism "stop-gain mutations" has arisen as a synonym for nonsense mutations. I avoid this locution because in my view it adds nothing but confusion.

***Frameshift Mutations:*** These are mutations that add or delete one (or two) base pairs. They are not substitution mutations, but they share many properties with point mutations, including their recombinational behavior in crosses and their ability to revert to the wild-type sequence.

Frameshift mutations have many very important additional properties. Because the amino acid code is read in triplets, these mutations result in a change in the reading frame, such that all the codons after the frameshift are read out of phase. As a consequence, all the subsequent codons are the wrong codons for the encoded polypeptide.

This example shows what happens to a hypothetical reading frame if a mutation causes an A to be inserted at the second codon after ATG:

Met His His His His His Asp Asp Stop

mRNA   5′ ... ATG CAT CAT CAT CAT CAT GAT GAT TAG ... 3′

Met His Thr Ser Ser Ser Stop

Insert A   5′ ... ATG CAT ACA TCA TCA TCA TGA TGA TTA G ... 3′

As with nonsense mutations, when a frameshift mutation occurs early in the coding sequence, the mutation is usually functionally null. The translation of the new reading frame after the mutation is usually terminated early because of the stop codons that are present at a substantial frequency (3 of 64, on average) in out-of-frame sequences. As detailed in Chapter 8, frameshift mutations can often be suppressed by mechanisms unique to them.

*Indels:* This is a term that is often used in sequence alignments to denote the addition or deletion of bases, usually of a very small number of bases, such as frameshift mutations.

## Mutations Involving More Than a Few Base Pairs

Not all mutations involve just one or a few base pairs. This became clear very early on in the history of genetics. The American geneticists Barbara McClintock, in her studies of maize chromosome mechanics, and Calvin Bridges and Alfred Sturtevant, in their studies of *Drosophila* polytene chromosomes, described several kinds of chromosomal rearrangements. All of these were consistent with the idea that genetic maps (based on recombination frequency in crosses) correspond with the order of physical sites on the chromosomes. These observations ultimately led to functional genes (as identified by inherited traits) being associated with cytological features along the chromosomes. Cytological analysis is still useful, even in the days of genome sequencing, because many inherited diseases (especially cancers) are characterized by uniquely placed chromosomal changes that involve millions of base pairs, which I describe below.

*Translocations:* Translocations are instances in which the arms of two different chromosomes become joined together (see Fig. 7.4), an event that is readily observable in the cells of most organisms, including in human cells, using cytogenetic techniques and light microscopy. When there is a simple exchange of chromosome arms between two nonhomologous chromosomes, the translocation is said to be "reciprocal." When no segments of genome are gained or lost in the event, the translocation is "balanced" and often confers no phenotype, although complications may arise during meiosis when the chromosomes must pair; this can sometimes result in unbalanced translocation in the offspring, in which chromosomal segments are gained or lost. Balanced reciprocal translocations are common in the human population: They are found in about 1 in 600 newborns.

However, translocations are a common cause of gain or loss of genomic DNA, and can cause a number of phenotypes. Sometimes the simple act of joining two chromosome segments together can result in abnormal gene regulation, even

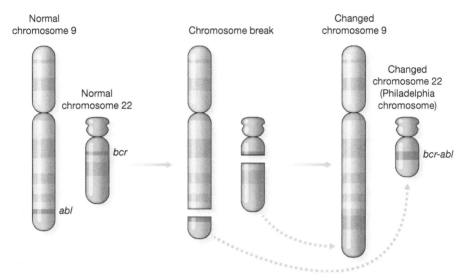

**Figure 7.4.** The Philadelphia chromosome (shown on *far right*) is a translocation between human chromosomes 9 and 22 that is associated with chronic myelogenous leukemia. It brings together two genes, *Bcr* (red) and *Abl* (blue), from each chromosome. (Redrawn, with permission, © 2007 Terese Winslow, U.S. Govt. has certain rights.)

when little or no sequence is gained or lost. Not surprisingly, therefore, translocations of many kinds are very common features of cancer, and in many cases can be shown to be functional in making the growth of cancer cells more vigorous.

For example, one of the first recurring chromosomal rearrangements associated with a human cancer is the so-called "Philadelphia chromosome," which is the result of a translocation between human chromosomes 9 and 16 (see Fig. 7.4). In this particular case, the breakage and rejoining event has regulatory consequences that drive the proliferation of the tumor cells in chronic myelogenous leukemia. The regulatory consequences of this translocation are discussed in more detail in Chapter 12.

*Copy Number Variants:* Changes in the number of copies of DNA segments can occur, ranging in size from a few thousand base pairs (enough to contain just one or a few genes) to large segments comprising substantial fractions of a chromosome arm. These changes are called "copy number variants" (or more commonly, CNVs), and most are too small to be seen cytogenetically and are only detectable using molecular methods. Smaller CNVs have recently come to prominence because of improvements in molecular techniques that have revealed their high prevalence. CNVs turn out to be surprisingly common in the genomes of all organisms examined, including those of humans.

Gene duplications have been known of since the early 20th century, when Alfred Sturtevant and Thomas Hunt Morgan inferred that a duplication on the *Drosophila* X chromosome results in the Bar-eye phenotype (which is characterized by a slit-like eye in place of the normal oval shape). A very common result of strong selection for the increased function of an expressed gene is its amplification. Amplification describes an increase in the number of copies of a gene without a proportional increase in the copy number of other genes. Such selective increases in copy number have been demonstrated in bacterial and yeast evolutionary studies. CNVs are common in tumor cells, with or without accompanying translocations. Many of the CNVs that are found in tumors have been associated with the amplification of genes that promote tumor growth.

**Insertion Mutations:** When functional genomic sequences are interrupted by relatively large, unrelated segments of DNA, the mutation is called an "insertion mutation." Such insertions can comprise thousands of base pairs, typically derived from a virus, a plasmid, or a type of mobile DNA element called a "transposon."

**Transposons:** Some specialized DNA segments have the ability to insert, at a low frequency, into DNA sequences more or less at random. These are called transposons. Some transposons encode the means of their own transposition; others are incomplete copies that are dependent on intact transposons in the same cell. Transposons are found in all kinds of organisms. They generally turn out to be the most frequent source of insertion mutations. A surprisingly large fraction of the DNA in higher eukaryotes consists of repeated and diverged copies of transposons or of their remnants, most of which have lost the ability to transpose.

The biology of transposons is fascinating, beginning with the prescient observations and brilliant experiments of Barbara McClintock in maize and other plants in the middle of the last century (well before the discovery of the DNA structure by Watson and Crick), many of them appreciated only decades later. The rediscovery of transposable elements in bacteria in the 1970s and then in virtually all other organisms enlightened us about, and has become central to our modern understanding of, the evolution of genomes. Transposons provide plausible mechanisms for how DNA segments might regularly be duplicated, amplified, and diverged, an essential feature of molecular evolutionary thought, about which I will have more to say later in this book.

Insertion mutations most commonly confer a null phenotype, usually because a coding sequence has been interrupted. However, there are situations in which the inserted viral or transposon DNA carries information that results in the activation of an otherwise unexpressed gene, resulting in a new dominant phenotype.

In humans, as in most other organisms, the most common types of insertion mutants are derived from transposons. The integration of viral sequences into an organism's genome has been found to cause the activation of genes, leading to cancer in some animals. In a few, unfortunate cases associated with the premature use of gene therapy, in which viruses were used to introduce a corrected gene into the cells of affected individuals, such viral insertions have also been associated with cancer in humans as well.

*Inversions:* Although they are relatively hard to detect, inversions are regularly observed in genomes from natural populations. Typically, they have no obvious phenotype, but sometimes they cause a gene to become inactive or to come under new regulation.

## DNA Recombination Mechanisms Contribute to Genomic Sequence Diversity

In their 1953 paper announcing the discovery of the double-helical structure of DNA, James Watson and Francis Crick famously wrote: "It has not escaped our notice that the specific pairing we have postulated immediately suggests a possible copying mechanism for the genetic material." But this was not the only important biological consequence that the structure "immediately suggests." Specifically, the structure of DNA provides a simple explanation for the generation of point mutations: rare errors made by the copying mechanism. It also provides an explanation for why it is that recombination occurs only where parental chromosomes share sequence identity (often referred to as "sequence homology"). This is because the mechanisms that recognize sequence identity do not prevent recombination from occurring where a low level of diversity in sequence ($\sim$1% or less) exists. Analogous to the generation of point mutations by rare errors in copying, the generation of translocations would be the result of rare errors made by these recombination mechanisms.

It turns out that there are quite a few DNA recombination and repair mechanisms, each of which can contribute to the generation of diversity. For example, when DNA is damaged and requires repair, the repair mechanisms sometimes make errors, resulting in mutations. A detailed discussion of all these mechanisms and the way in which they can contribute to diversity is beyond the scope of this book, but a few deserve an explicit mention.

*Homologous (or Legitimate) Recombination:* This phrase refers to the occurrence of recombination between essentially identical, long stretches of genome—for example, between paired homologous chromosomes in a eukaryote during meiosis, as described in Chapter 3. When recombination between paired homologous chromosomes takes place, genetic material is exchanged between them in a process also called crossing over. In a population, homologous recombination during meiosis generates diversity by allowing a mutation that occurred in one lineage to recombine with another mutation in another lineage, thus generating a double-mutant lineage in that population. This appears to be very important for all organisms: Even haploid bacteria and viruses have evolved ways to exchange genome segments by legitimate recombination.

Homologous recombination can take place during mitotic growth, as well as during meiosis. The term is used to refer to recombination between sequences that contain only short stretches of homology, much shorter than a chromosome arm. This can occur via several diverse mechanisms, including by DNA repair processes or as a result of rare mitotic recombination events between homologous (or partly homologous) repeated segments in which chromosomes are misaligned. These kinds of recombination events are very important in our thinking about evolutionary process because they can account for the amplification, deletion, and fusion of sequences that produce new genetic variants. These variants can sometimes provide sufficient improvements to an organism's survival and reproductive fitness to cause their retention by selection. Recombination events of this kind are engines that drive genomic diversity, leading to evolutionary change, as I describe in Chapter 13.

Homologous recombination is tightly regulated and largely limited to meiosis, especially in organisms that contain substantial amounts of repeated DNA. Nevertheless, homologous recombination mechanisms contribute to the generation of deletions and duplications, especially in regions of the genome that are flanked by repeated sequences, such as transposons or their remnants. To see the importance of homologous recombination between repeated DNA sequences (notably between transposons and their remnants), see Figure 7.5, which shows how homologous recombination within these sequences when in the same orientation (sometimes referred to as "direct" repetitions) can produce deletions, together with a reciprocal circular by-product, which is usually, but not always, lost. This illustration also shows how recombination within repeated sequences in the opposite orientation (sometimes referred to as "inverted" repetitions) will produce inversions. The inversion event is clearly reversible by homologous recombination. However, the deletion event is reversible if, and only if, the circular product is somehow retained. It is thought that such circular products are sometimes retained by selection, and if they find a way to recombine with another homologous repeated sequence at another position, the result can lead to amplifications of genomic sequences, such as those found in many cancers.

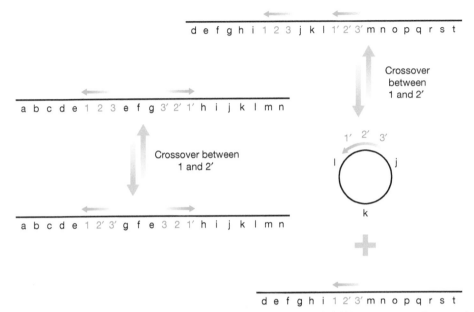

**Figure 7.5.** Consequences of recombination within repeated DNA sequences. Repeated sequences (1, 2, 3) are shown in red, and apostrophes highlight the fate of the two repeated copies after recombination. (*Left*) Crossover between repeated sequences in opposite orientation, leading to an inversion. (*Right*) Crossover between repeated sequences in the same orientation, leading to the excision of intervening sequence and its formation into a circular product.

***Mitotic Recombination:*** As mentioned above, homologous recombination usually occurs during meiosis but can also occur between intact homologous chromosomes in cells that reproduce mitotically and are not undergoing meiosis. This mechanism—mitotic recombination—is a common cause, for example, of the loss of a functional allele in animals that are heterozygous for a "recessive oncogene" (sometimes called a "tumor-suppressor" gene, a term I try to avoid but which has become standard). After the crossover event, recombinant cell lineages are generated, one of which has no functional allele of the oncogene and goes on to give rise to tumor cells.

***Illegitimate Recombination:*** This is a problematic term that is used by some geneticists for any recombination event that involves less than a chromosome arm of homology. I use the term only for events that really involve no homology at all, such as transposition or "nonhomologous end joining," which happens when free DNA ends are simply fused together. I include in the category of homologous recombination events that occur within regions of sequence similarity of the order of 100 bases because they invariably turn out to be catalyzed by enzymes identical, or very similar, to the ones that catalyze "legitimate" recombination.

*Transposition:* When a transposon integrates into a new site in a genome, the process is called "transposition," and it involves no homology between the transposon and the site of insertion. The result of this integration is an insertion mutation and usually the addition of a copy of the transposon into the genome. It is thought that this feature of transposition allows transposon sequences to accumulate, accounting for the observation that much of the repeated DNA in many organisms is transposon derived. As indicated above, intact active transposons encode an enzyme that catalyzes their insertion. However, most copies of transposons in the genome are not active, and as a result, their sequences have become diversified by mutation and drift over evolutionary time.

*Gene Conversion:* Gene conversion describes a process whereby one copy of a gene in an organism is replaced by a copy of a homolog as a result of the DNA-copying mechanisms associated with DNA repair or recombination.

All of the above mechanisms will produce, in a population, new genotypes that can be subject to selection, based on their contribution to fitness. Readers interested in recombination mechanisms per se will find an abundant literature about them, and extensive coverage in standard genetics textbooks.

For the purposes of this book, it suffices to summarize the role of recombination and repair mechanisms and their frequency:

- *One in 10,000 events or less per average-size gene per generation:* Mutations that cause deletion, duplication, or inversion by illegitimate recombination are sometimes more frequent than point mutations, which occur in humans at about $1/100,000$ per average-size gene per generation. In humans, new transposon insertions are less frequent than point mutations.

- *Once per generation per chromosome:* Homologous recombination during meiosis is the most frequent and important way in which existing DNA sequences are rearranged in a population. In humans (as in most higher eukaryotes), this occurs about once per chromosome arm per generation.

- *Once every 10 to 100 generations per chromosome:* Duplications, deletions, and inversions arising from homologous recombination between relatively short regions of sequence homology, such as between individual genes or exons, transposons or transposon remnants, or highly repeated simple sequences, occur at this frequency. Mitotic recombination events and gene conversions, which both serve to reduce diversity among repeated sequences, can occur at this kind of frequency as well.

## INTRODUCTORY BIOGRAPHIES

**Barbara McClintock (1902–1992)** was an American geneticist who began as a pioneer in cytogenetics. She was legendary for her skill in visualizing chromosomes. She used this skill to show directly in maize that the physical crossing over she could see in cytological preparations was correlated with the recombination of genetic markers on the chromosome. She rose to scientific eminence early in her career: She was elected to the National Academy in 1944 and elected President of the Genetics Society in 1945. In later years, she discovered and completely elucidated genetically the phenomenon of DNA transposition and the regulatory changes that accompany the insertion or removal of transposons, for which she was awarded a solo Nobel Prize.

**Calvin Blackman Bridges (1889–1938)** was an American geneticist who obtained his PhD working with Thomas Hunt Morgan and Alfred Sturtevant in the Columbia fly room and moved with them to Caltech, where he spent the rest of his career. It was he, more than anyone else, who established and exploited the banding pattern in the polytene chromosomes of *Drosophila*. He showed the correlation between this pattern and the genetic consequences of genomic rearrangements of all kinds, including deletions, duplications, translocations, and inversions.

**Francis H.C. Crick (1916–2004)** was a British biophysicist and molecular biologist. He was, in many ways, the leading theorist of molecular biology. With James D. Watson, he discovered the double-helical structure of DNA, based on the theory of helical crystallography, to which he contributed significantly. With Sydney Brenner, he contributed to the elucidation of the genetic code and contributed the now standard codon table. He put forward the adaptor hypothesis (i.e., that translation required a molecule with an anticodon at one end and an amino acid at the other); this turned out to exactly describe tRNA.

**Walter Gilbert (b. 1932)** is an American physicist turned molecular biologist who played a leading role in the founding of the both the biotechnology industry and the science of genomics. Gilbert was the first to isolate and purify a regulatory protein (the lac repressor) and study its properties. He invented an early method for sequencing DNA and used it to find the sequences of regulatory sites that control the lac operon. He was among the first to use recombinant DNA technology to produce human proteins of commercial value in bacteria, and he became a founder of one of the first successful biotechnology companies. He was an early advocate for sequencing the human genome and anticipated the transformation of molecular biology into and information science through the availability of molecular sequences. He introduced the words "intron" and "exon" in a prescient article that proposed a rationale for their evolution.

**James D. Watson (b. 1928)** is an American molecular biologist who, with Francis Crick, discovered the double-helical structure of DNA. He began, in the 1960s, to write a textbook, *The Molecular Biology of the Gene*, that has served to introduce students to and to define the subject ever since. He also wrote a popular best-selling account of the DNA structure discovery, *The Double Helix*, that made him an international celebrity beyond the boundaries of science.

# CHAPTER 8

# Suppression Genetics

## Introduction

Most mutations, especially those that result in a simple recessive (loss-of-function) phenotype, consist of simple changes in DNA sequence, most commonly single base-pair substitutions. Generally, the consequences of such mutations are deleterious, especially when they occur in sequences that encode proteins. Such mutations can revert to restore the wild-type phenotype by the conceptually simple mechanism of back mutation, in which a mutant base pair changes back again to restore the original DNA sequence. Such a reversion will result, of course, in loss of any phenotype caused by the original mutation. The frequency of such events is rare, generally of the same order as of the forward mutation rate.

*True Revertants:* These revertants no longer have the mutant phenotype because the alteration in DNA sequence that caused the mutant phenotype has been reversed. They simply return to the parental genotype and phenotype.

*Pseudo-Revertants:* In circumstances under which one can select for restoration of a loss-of-function phenotype, one recovers, in addition to a true revertant, another kind of revertant, in which the original mutation remains and another mutation, at a different site, somehow compensates for the consequences of the original mutation. These are called "pseudo-revertants." Although pseudo-revertants no longer have the mutant phenotype, they retain the alteration in DNA sequence that caused the mutant phenotype.

*Suppression:* This is a general term for the many different ways in which a mutation at a second site can reverse the phenotype of a mutation. Geneticists call the second-site mutation in such a pseudo-revertant a "suppressor mutation." Suppression is a very general phenomenon, and can happen in many different ways. There are well-documented examples of suppression in every organism that has had its genetics studied, including humans. However, much of what we know about the many mechanisms of suppression has come from studies of single-celled microorganisms, such as bacteria or yeasts, which are especially convenient for studies of mutation and reversion. Before I discuss the findings of such studies in more detail, there are two more terms to bring to the reader's attention.

*Intragenic Suppressors:* Second-site mutations in the same cistron that reverse the phenotype of the original mutation are called "intragenic suppressors."

*Extragenic Suppressors:* Second-site mutations in a different functional gene that somehow reverse the phenotype of the original mutation are called "extragenic suppressors." Often, these are completely unlinked to the original mutation. The mechanisms of action of extragenic suppressors are extremely diverse, ranging from systematic alterations in the way codons are read to changes to specific functional interactions among genes and their products. It is usual to distinguish suppressors on this basis.

## Genetic Properties of Suppressors

How can one distinguish between true revertants and pseudo-revertants? In principle, one could sequence the mutant and revertant genomes. If the mutation is still present in the revertant and there is, somewhere else in the genome, another plausible mutation, we would have a prima facie argument for suppression, although not always a rigorous one. This method, however, is laborious and expensive compared to genetic analysis, which, as we shall see, can be much more informative and rigorous than the simple detection of suppression. Also, as is common to all "just sequence" approaches, the mutation rate in most organisms is high enough that one is likely to find other, irrelevant, random base changes that do not contribute to the phenotype. Genetic analysis would then be required in any case to make the case that the plausible second mutation is actually the cause of the suppression.

I have chosen to use microbial systems as the examples for the genetics of suppression. Much of the genetics of bacteria and fungi was worked out using nutritional mutations in microorganisms because they are such excellent and convenient genetic markers. In bacteria and haploid fungi, phenotypes directly reflect genotypes. One can readily combine, by recombination, mutations in different biosynthetic pathways and follow the mutations in crosses, and also use other manipulations, simply by following which nutrients are required for growth.

As discussed in Chapter 4, auxotrophs are mutants that lack metabolites normally synthesized by the organism; thus, they can only grow when fed these metabolites. True prototrophs are species that can grow on just sources of carbon (C), nitrogen (N), sulfur (S), phosphorus (P), and salts. The cognate wild-type phenotype to auxotrophy is prototrophy. However, over time, the use of these words by geneticists has come to include statements such as "Strain X is auxotrophic for arginine, but prototrophic for lysine." This is convenient, but not quite rigorous, because there are very many kinds of microorganisms (including many human pathogens) that in the wild have elaborate metabolic requirements, in addition to requiring sources of C, N, S, P, and salts. Auxotrophy in these cases is used nevertheless to refer to any additional metabolic requirements that result from mutation. The usage of "prototrophic" for the cognate wild-type phenotype in these cases should therefore be undertaken with care.

For studies of suppression genetics, the important property of auxotrophic mutations is that they are a kind of conditional-lethal mutation, the phenotype of which is easily scored and yet specific to the gene in which the mutation lies. The "condition" here is the presence or absence of a particular added metabolite; the permissive condition is the presence of the metabolite, and the nonpermissive condition is its absence. As in the T4 $rII$ system, one can select against auxotrophic mutations by simply leaving the required metabolite out of the growth medium, just as one can select against bacteriophage T4 $rII$ mutations by requiring growth on K12($\lambda$) (as explained in Chapter 6). A crucial advantage of bacteria and yeasts is that in a strain with several auxotrophies, one can select against each of them either individually (by leaving out a single metabolite) or in groups (by leaving out several).

Yeast auxotrophs are the first example I turn to, because yeasts can be grown as haploid cultures and, as mentioned before, recessive phenotypes can be detected and followed without any further manipulation. Moreover, in yeasts, unlike in bacteria, analysis by Mendelian genetics is straightforward. Historically, it was just such an analysis that led to the discovery of nonsense suppression in yeasts by the Canadian biophysicist Robert Mortimer and the America geneticist Donald Hawthorne in the early 1960s.

Consider a recessive auxotrophic mutant that requires tryptophan, containing a mutation we will call $trp^-$. It is easy to select for revertants of this mutation simply by looking for colonies on culture plates made with an agar medium that lacks tryptophan. We will assume that we have recovered revertants from independent cultures to avoid duplication. In principle, as we have indicated, the tryptophan-independent phenotype could be the result of a true reversion (i.e., a reversal of whatever DNA sequence change had occurred to produce the $trp^-$ mutation). In this case, the genotype of our revertant would be $trp^+$—that is, exactly what it was before the auxotrophic mutation occurred. Alternatively, the tryptophan-independent phenotype could be the result of a second mutation, a suppressor. In this case, it is a pseudo-revertant with a genotype we will denote as $trp^- sup$. Lack of a suppressor we will denote as $nosup$. This notation is a little nonstandard, but because the notations for suppressors in yeast and bacteria are very different and not always adhered to, I adopted a simple, intuitive alternative for this book.

A simple, implicit experiment distinguishes true revertants from pseudo-revertants. Compare a cross between the true revertant (genotype $trp^+ nosup$) to wild type (genotype $trp^+ nosup$) to a cross of a pseudo-revertant (genotype $trp^- sup$) to wild type (genotype $trp^+ nosup$). Note that both of these crosses are between two strains that are phenotypically the same: All the parents grow in the absence of tryptophan. The way we distinguish a true revertant from a pseudo-revertant is as follows.

- In the case of a true revertant, the cross to wild type will produce no progeny that require tryptophan: The two genotypes are identical.

- In the case of a pseudo-revertant, the cross will produce recombinants between $trp$ and $sup$ and their homologous loci. The recombinant progeny that inherit $trp^-$ but not $sup$ from the pseudo-revertant parent will once again require tryptophan. Their reappearance in the cross is the genetic hallmark of suppression.

The doubly heterozygous diploid in the cross of a pseudo-revertant with wild type will be

$$\frac{trp^-}{trp^+} \quad \frac{sup}{nosup}.$$

The recombinant gametes will be $trp^- nosup$ and $trp^+ sup$, and the parental gametes will be $trp^- sup$ and $trp^+ nosup$. Because the gametes of yeasts produce haploid strains, the phenotypes of which directly reflect their genotypes, the classification of these recombinants is simple. Only one ($trp^- nosup$) of the four possible gametes will have the mutant requirement for tryptophan. If the $trp$ and $sup$ loci are unlinked, then the frequency of tryptophan-requiring progeny will be 25%.

If, on the other hand, the $trp$ and $sup$ loci are linked, then the resolution of this method will decline. In practice, this experiment works extremely well for the detection of suppression in yeast because the rate of recombination in this organism is very high—of the order of 1% per gene—meaning that one only needs to examine a few hundred meiotic products to detect a suppressor in an adjacent gene.

▶ *Genetic test for suppression: Cross phenotypic revertants to wild type and screen progeny gametes for the reappearance of the original mutant phenotype.*

Suppression is a very general phenomenon in genetics. The mechanisms by which the function of a mutant gene can be restored by a second mutation are very diverse. As mentioned above, they include two very important general classes: informational suppressors and functional suppressors. The former played a critical role in the development of our understanding of molecular biology, while the latter continue to play a major role in our understanding of gene interactions in all organisms.

**Informational Suppressor:** This class of suppressor restores function to mutations by altering some part of the process that translates DNA sequence into polypeptide chains. These suppressors work on particular kinds of mutations that can occur in virtually all genes.

Informational suppressors act to produce functional products from the gene affected by the mutation ($trp^-$ in the example above) by changing the information flow such that this mutation (and similar mutations in other genes) is reinterpreted by the translational machinery.

▶ *Informational suppressors should be mutation-specific and gene-nonspecific.*

**Functional Suppressor:** This class of suppressor changes the defective function in the mutant in a way that somehow compensates for the original mutant phenotype.

Functional suppressors alter the biology of the organism in some way that gets around the defective phenotype of the mutation (in the earlier example, the supply of tryptophan). Generally, there are many (sometimes all) mutations in a gene that can be suppressed by the same functional suppressor. As I discuss in Chapter 9, the ways in which this can happen are very diverse, and functional suppressors are a subset of mutations that alter gene interactions, the subject of Chapter 10.

▶ *Functional suppressors should be gene-specific and mutation-nonspecific.*

**Nonsense Suppressor:** A paradigm example of informational suppression is a nonsense suppressor. These suppressors cause the occasional insertion of an amino acid at mutations in stop codons that would otherwise have prematurely terminated translation. They come in three types that correspond to the three stop or termination codons (as indicated in Chapter 7); these codons do not specify an amino acid, hence the name "nonsense." One suppressor type affects TAG nonsense mutations (sometimes referred to as "amber" mutations for historical reasons); another affects TAA nonsense mutations (sometimes referred to as "ochre" mutations); and the third affects TGA mutations (these have no agreed-upon historical name; "opal" and "umber" are sometimes used). The mechanism of nonsense suppression is simple in concept: They are tRNA genes that have been altered by a mutation in the anticodon so that they bind a nonsense codon instead of the codon they normally recognize during translation. They are otherwise intact and functional adaptor molecules, and retain the ability to be charged with a particular amino acid and to donate it to the growing peptide chain during translation.

For example, a tRNA that normally recognizes CAG and inserts glutamine during translation can have its anticodon changed by a substitution mutation so that it now recognizes TAG instead of CAG. This suppressor mutant tRNA molecule still gets attached normally to glutamine by the amino acid–activating enzyme, but during translation this glutamine will now be inserted at TAG and not at CAG codons. Because most organisms have several tRNA genes for each codon, there are plenty of other, unmutated glutamine-tRNAs to insert glutamine normally at CAG codons. If there are, in the same strain, nonsense mutations that have changed a sense codon to a stop TAG codon (the codon table in Fig. 7.3 shows that many different codons can be changed to TAG by a base substitution mutation), the presence of the suppressor will restore a significant fraction of functional polypeptide synthesis to an otherwise null TAG chain-terminating mutant. Typical nonsense suppressors in bacteria and yeasts can restore around 20% of normal protein levels. Because most auxotrophic mutations cause loss of an enzymatic function, this level of restoration is more than sufficient in practice. Nonsense mutations in metabolic pathways are generally recessive (without being leaky), and suppression is often efficient enough to restore wild-type levels of growth in the absence of the required metabolite.

However, as I have mentioned, there are many different codons that are one change away from TAG: In addition to CAG (glutamine), one can get to TAG from AAG (lysine), GAG (glutamate), TAC (tyrosine), TAT (tyrosine), TCG (serine), TTG (leucine), and TGG (tryptophan). If one has a TAG suppressor that inserts glutamine, suppression will result, in most cases where glutamine is substituted for the amino acid encoded by the original, unmutated codon. The net effect will be a protein that has a different amino acid to that originally encoded. Most proteins are pretty robust to single missense mutations, so suppression by glutamine-inserting suppressor tRNAs generally works even for serine (TCG)-derived TAG mutants. However, one can sometimes observe better restoration of function by suppressors that replace the originally encoded amino acid (i.e., serine, in this example). This is a nice illustration, nevertheless, of the generalization that most missense mutations may not cause a detectable phenotype. It also nicely demonstrates why nonsense mutations are attractive to experimentalists because of their close approximation to a null phenotype, which is generally not "leaky."

Some of the genetic features of nonsense suppression are effectively illustrated by a historical anecdote from the earliest days of yeast molecular genetics. I have taken some liberties with the details in the interest of clarity and simplicity. Modern yeast genetics began with collections of auxotrophic mutations, which were sorted into functional genes by complementation tests and then used in recombination mapping. Just as Seymour Benzer favored bacteriophage T4 *rII* mutations with low growth in the nonpermissive host *E. coli* K12($\lambda$) (see Chapter 6), the yeast auxotrophic mutations that were most prized were those with the most reliable phenotype. That meant those with minimal growth background in the absence of the required nutrient and a normal growth rate in its presence. A series of well-behaved "standard" alleles, isolated and carefully characterized by Don Hawthorne and Robert Mortimer, became popular in the field. What made them well behaved, of course, was that they had the null phenotype—the complete absence of function of an enzyme required to make a metabolite.

Mortimer and Hawthorne constructed, by recombination, strains that contained many different auxotrophic mutations, each requiring a different metabolite. A widely used quintuple mutant required tryptophan, tyrosine, leucine, **adenine**, and lysine for growth (genotype *trp tyr leu ade lys*). All five of these genes turned out to be unlinked to each other by recombination.

Three remarkable properties of this strain emerged when revertants were studied.

1.  When revertants for the tryptophan requirement were selected, many turned out to be pseudo-revertants when crossed with wild type. Tryptophan auxotrophs were recovered at high frequency ($\sim$25%) from the cross between the two phenotypically tryptophan-independent parents. This simple cross thus revealed suppressor mutations that were, in each case, unlinked to the original mutation.

2.  More remarkable, virtually all the tryptophan-independent pseudo-revertants also lost their requirement for tyrosine and lysine! The true revertants (i.e., those that when crossed with wild type yielded no tryptophan-independent progeny) did not have this property. Thus, the suppressors suppressed not just one but all three of these requirements. But they always failed to suppress the need for adenine or leucine.

3.  When revertants for the adenine or leucine requirements were selected, once again there were many pseudo-revertants with unlinked suppressor mutations. And again, the requirements for adenine and leucine were both suppressed by suppressors selected for the loss of either, whereas the other three mutations remained unsuppressed.

By now, the reader will have realized that the *trp*, *tyr*, and *lys* mutations might have been one kind of nonsense mutation (in this case TAA, or ochre) and the *leu* and *ade* must have been another kind (in this case TAG, or amber). By favoring null phenotypes, Mortimer and Hawthorne had biased their choice of "well-behaved" alleles toward nonsense mutations and away from missense mutations, just as Benzer, by preferring very low rates of reversion, had biased his choice toward alleles that turned out to be deletions in *rII*.

The yeast auxotroph anecdote illustrates a major feature of informational suppressors: They are mutation-specific (i.e., TAA suppressors suppress TAA mutations and not TAG or other kinds of mutations), but they are gene-nonspecific. In general, TAA suppressors will suppress virtually any mutation in any gene that has a TAA codon at the site of the original mutation. Exceptions to this rule are uncommon but not all that rare, and most are readily understood: Each of the suppressors inserts an amino acid, which could be (consulting the codon table in Fig. 7.3) Gln, Leu, or Tyr. Occasionally, the substitution of one amino acid (in our example, Gln) for what might, in the wild-type protein, have been another (Leu or Tyr) that was mutated to TAA does not work in the encoded protein.

▶ *Nonsense suppressors are allele-specific and gene-nonspecific.*

Finally, nonsense suppressors are usually dominant, as one might expect. A diploid strain heterozygous for the suppressor gene and homozygous for an auxotrophic nonsense mutation will grow in the absence of the required metabolite in the medium.

## Mutual Suppression by Frameshift Mutations

In the previous discussion in Chapter 7 of different classes of mutations, I provided a hypothetical example of a simple frameshift:

```
Wild-type       Met His His  His His His Asp Asp Stop
mRNA      5'... ATG CAT CAT  CAT CAT CAT GAT GAT TAG...3'
                Met His Thr  Ser Ser Ser Stop
Insert A  5'... ATG CAT ACA TCA TCA TCA TGA TGA TTA G...3'
```

The main point of this example is that everything that ribosomes translate after the frameshift is gibberish and eventually chain-terminating. This is because the ribosomes are now translating in a new reading frame, which is called the +1 reading frame.

Now consider the possibilities for reversion. True revertants, of course, would feature the removal of the original mutation, the A inserted after the second codon. However, the removal of another base in the same codon might result in the restoration of the correct reading frame, which would result in a product that has only a single amino acid substitution, very possibly with no phenotype.

There are other possibilities. The most illuminating and important of these are mutations that restore the reading frame by a second mutation, such as deleting T in the fifth codon:

Wild-type     Met His His His His His Asp Asp Stop
mRNA     $5'$...ATG CAT CAT CAT CAT CAT GAT GAT TAG...$3'$

     Met His Thr Ser Ser Ser Stop
Insert A   $5'$...ATG CAT ACA TCA TCA TCA TGA TGA TTA G...$3'$

     Met His Thr Ser His His Asp Asp Stop
Delete T   $5'$...ATG CAT ACA TCA CAT CAT GAT GAT TAG...$3'$

This second mutation, by definition a suppressor mutation, restores the original reading frame with the exception of the codons between the insertion of A and the deletion of T. The resulting amino acid substitutions may or may not cause the loss or alteration of protein function. Frequently, relatively short alterations of this kind have minimal or undetectable phenotypic consequences.

Unlike nonsense suppression, there is a symmetry in frameshift suppression: The mutation and the suppressor both shift the reading frame, and each by itself has similarly disastrous consequences. Specifically, if in our example one were to remove the original mutation, one would once again shift the reading frame:

     Met His Thr Ser His His Asp Asp Stop
Delete T   $5'$...ATG CAT ACA TCA CAT CAT GAT GAT TAG...$3'$

     Met His His His Ile Met Met Ile
Delete A   $5'$...ATG CAT CAT CAC ATC ATG ATG ATT AG...$3'$

Once again, everything beyond the codon from which we deleted a base is gibberish because of a shift in reading frame, but this is a different reading frame from the one to which the original translation was shifted by the insertion of A. This reading frame is called the −1 reading frame.

Frameshift mutations are often described in terms of the reading frame to which they shift translation. Thus, an insertion of one base causes a +1 frame-shift, and the deletion of a base causes a −1 frameshift. Because there are only three possible reading frames, the insertion of two bases also causes a −1 frame-shift, and deletion of two bases causes a +1 frameshift. It should be clear that the addition or deletion of three bases leaves the reading frame intact, although the missing or added amino acid might well cause a phenotype. It was the inference, in 1961, of these genetic properties by Francis Crick and the British geneticist Sydney Brenner that provided the first clear evidence that DNA sequence is decoded as triplet codons.[1]

Crick and Brenner used Benzer's *rII* system in bacteriophages (see Chapter 6) to infer the genetic properties of frameshift mutations. Their discovery of mutual suppression by +1 and −1 frameshift mutations came just before the biochemical elucidation of the genetic code. Brenner and Crick's work was a brilliant example of genetic analysis, done long before DNA sequencing was possible. Instead, they exploited Benzer's fine-structure genetics of the T4 *rII* locus. They selected rever-tants of suspected frameshift mutations in *rII*, crossed them back to wild type, and recovered and mapped both the individual frameshift mutations from pseudo-revertants. They then used Benzer's mutants (especially the deletions) to do the fine-structure mapping.

Crick and Brenner took special advantage of the deletions that spanned the end of the A cistron and the beginning of the B cistron, which removed the boun-dary between them. These deletions fused the A and B cistrons together into a single hybrid cistron. The translation of this hybrid cistron begins at the start of *rIIA* and continues right into the start of the *rIIB* cistron without interruption. The fused protein has rIIB function, but not rIIA function. In the background of these deletions, they could select and assess rIIB function from suppressed frameshifts that contained considerable stretches of DNA that were translated out of frame, without producing phenotypic consequences on rIIB function.

Finally, I should highlight that the mutual suppression of +1 and −1 frame-shifts works only in the *cis* configuration in the *cis–trans* test. Formally, then, the suppression phenotype of frameshift mutations would be classified as *cis*-dominant, whereas the functional phenotype is almost always recessive, because frameshifts most commonly cause complete loss of function. The concept of *cis*-dominance is a cornerstone of genetic analysis of gene regulation, to which I will return in Chapter 11.

---

[1] Crick FH, Barnett L, Brenner S, Watts-Tobin RJ. 1961. General nature of the genetic code for proteins. *Nature* **192:** 1227–1232.

*tRNA-Based Frameshift Suppressors:* These suppressors work by lengthening their anticodon from three to four bases. In this way, they can cause ribosomes to shift their reading frame in the $-1$ direction and thereby suppress the effect of a $+1$ frameshift nearby. These suppressors, like nonsense suppressors, are simply dominant with respect to their suppression phenotype because a heterozygous diploid still has the suppression phenotype. They confirm that it is tRNA binding, not the ribosome per se, that maintains the reading frame during translation.

*Ribosomal Ambiguity Suppressors:* Ribosomal ambiguity suppressors suppress many kinds of mutations (including missense mutations) by reducing the specificity with which tRNAs are matched with their cognate codons in the ribosomes. They usually occur as mutations that alter a protein constituent of the ribosome. Their phenotypes and dominance relationships can be complex.

## INTRODUCTORY BIOGRAPHIES

**Robert K. Mortimer (1927–2007)** was a Canadian biophysicist and geneticist who spent his career working on the genetics of budding yeasts. His first major contribution was to find a practical way to dissect the asci (the products of meiosis), making tetrad analysis possible at a very large scale. For decades, he organized and edited the recombination map for the entire worldwide yeast community, which produced a spirit of sharing that did much to catapult the yeast system to the forefront of microbial genetics.

**Donald Hawthorne (1926–2003)** was an American yeast geneticist who spent his career working on the genetics of yeast. He had begun mapping mutations when Mortimer's tetrad dissection method became available, and the two joined forces to make the first comprehensive genetic map. His lasting contributions include the genetic analysis of the galactose utilization genes of yeast. He also made major contributions to the understanding of the control of mating type.

**Sydney Brenner (b. 1927)** is a South African–born British geneticist and molecular biologist. He was central to elucidating the genetic code and, especially, the analysis of tRNA nonsense suppressors and reciprocal frameshift suppression using phage genetics. In later years, he established the nematode *Caenorhabditis elegans* as a model organism for genetic studies of development and neurobiology, which earned him and his colleagues a Nobel Prize.

CHAPTER 9

# Functional Suppression

There are many different kinds of suppressors that reverse mutant phenotypes. Unlike informational suppressors, functional suppressors do not affect the transfer of information from the DNA of a gene to its molecular product. Instead, via a mutation in a second gene, they compensate for a defect caused by the original mutation. These types of second-site mutations are too many to enumerate here, so I have chosen to describe only a few examples that I think will be the most illustrative. All have in common that they reverse the phenotypes of many kinds of mutations in genes involved in a particular biological function or pathway, as opposed to particular types of mutations in diverse functions or pathways.

> ▶ *Functional suppressors are gene-specific or pathway-specific, and not limited to the suppression of particular kinds of mutations.*

Functional suppressors are of general interest to geneticists because they provide a path to inferring interactions between genes. As we will see, experimental geneticists have made good use of deliberate screens for suppressors in order to discover and understand these interactions. Functional suppressors are the most intuitive of the many phenomena that geneticists can study that involve gene–gene interactions, the subject of Chapter 10.

## Protein Interaction Suppressors

Some suppressors compensate for a loss of mutant protein function via mutations that affect a second protein that interacts with it—these are called "protein interaction suppressors." It is not at all surprising to find such protein–protein relationships in complicated biological systems that consist of many different polypeptides that bind together to provide function. Examples of systems that involve interactions among many protein subunits are abundant in the genetic literature: the assembly of virus particles, the function of internal cellular structures (such as the cytoskeleton), the structure and function of the DNA replication apparatus, and the spindles that separate chromosomes during mitosis and meiosis. Indeed, one could imagine finding such compensating mutations wherever protein interactions are important. I have chosen a few historical examples that illustrate this kind of genetic interaction in model systems, because I believe that all of these are likely to be recapitulated in any organism, including humans.

Among the earliest and clearest examples of interaction suppression is the cooperation of viral and host proteins during lytic growth of bacteriophage λ in its normal host, the bacterium *Escherichia coli*. In a landmark paper[1] in 1971, the Greek-born American microbiologist and biochemist Costa Georgopoulos and the American geneticist Ira Herskowitz isolated a series of mutants, specifically carrying mutations in the *E. coli* genome but not in the λ genome, that could not support the growth of λ but that otherwise produced no *E. coli* phenotype. To avoid a tangle of ancient nomenclature, I will call them "inhospitable" host mutations. Georgopoulos and Herskowitz then found λ phage mutants (carrying mutations in the λ genome, but not in the *E. coli* genome) that had recovered the ability to grow on the mutant hosts. These absolutely had to be suppressors and not true revertants of the inhospitable host mutations because the mutations that reversed the inhospitable phenotype were found in a different genome—that of the λ phage. The acquisition of the suppressor phenotype was not accompanied by any growth disadvantage to the phage.

Genetic mapping studies showed that the inhospitable host mutations that restricted the growth of λ occurred in at least three different *E. coli* genes (I will call them *A*, *B*, and *C*, for clarity), each distant from the others on the *E. coli* genome. Two of these genes encode proteins that had pretty clear functions, even then. Gene *A* encodes a subunit of the complex of proteins that replicate the *E. coli* bacterial DNA—its true name is *dnaB*. Gene *B* encodes a subunit of the *E. coli* RNA polymerase, responsible for the transcription of DNA into RNA molecules of various kinds—its true name is *rpoB*. Gene *C*, characterized only much later, encodes in *E. coli* a chaperone protein that aids in the folding of other proteins—its true name remains *groE*, the name given to it by Georgopoulos and Herskowitz. The suppressor mutations that restore to λ the ability to grow in the inhospitable hosts were in three different phage genes,

---

[1] Georgopoulos CP, Herskowitz I. 1971. *Escherichia coli* mutants blocked in lambda DNA synthesis. In *The bacteriophage lambda* (ed. Hershey AD), pp. 553–564. Cold Spring Harbor Laboratory Press, Cold Spring Harbor, NY.

each of which had been previously characterized with respect to their viral function. One of these ($\lambda$ gene $P$) was implicated in phage DNA replication: Null mutants of this gene fail to replicate phage DNA. The second ($\lambda$ gene $N$) was implicated in transcription: Null mutants fail to transcribe most of the phage genome after infection. The third ($\lambda$ gene $E$) encodes the major subunit of the viral capsid.

Many readers may have anticipated by now how this turned out to be such a clear illustration of functional suppression. The suppressors of the inhospitable mutations in bacterial DNA replication gene $A$ occur in $\lambda$ gene $P$, the gene involved in DNA replication in the phage; the suppressors of mutations in gene $B$ (encoding an RNA polymerase subunit) occur in $\lambda$ gene $N$, the gene involved in transcription of the phage genome; and the suppressors of mutations in gene $C$ (involved in protein folding) occur in $\lambda$ gene $E$, the major subunit of the capsid that must be coaxed into a polyhedral shell in which many subunits occupy not-quite-identical positions, a difficult protein-folding problem.

Beyond the nice correspondence of phenotypes and functions, over time biochemical analysis fully confirmed that the product of host gene $A$ binds to the protein encoded by $\lambda$ gene $P$ in order to successfully replicate $\lambda$ DNA. The bacterial RNA polymerase subunit that is the product of host gene $B$ has to bind the protein encoded by $\lambda$ gene $N$ to extend transcription across the entire phage genome. Finally, the product of host gene $C$ must bind nascent E protein to form a functional phage capsid.

## Suppressors with Novel Phenotypes

Functional suppression provided experimental geneticists with a conceptually simple route to the discovery of interacting partners among genes and proteins. If one has a mutation that affects a function of interest, functional suppression is a way in which to find other genes that affect the same function or a closely related one. In the 1970s, this idea was taken up by several laboratories interested in phage morphogenesis, including my own. The assembly of a phage is a process that already then was known to involve many proteins, but only a few were well characterized. Very little was known, for example, about how DNA was encapsulated into the protein shell or how the assembled phage injected DNA into a new host.

American geneticist and cell biologist Jonathan Jarvik studied revertants of temperature-sensitive—heat-sensitive (Ts) or cold-sensitive (Cs)—conditional-lethal phage mutations, some of which he expected might contain functional suppressor mutations in genes encoding interacting proteins.[2] Because a single point mutation can, as emphasized in the earlier discussion of human hemoglobin (see Chapter 2), have several phenotypes, Jarvik screened many independent stocks of Ts morphogenesis mutants for spontaneous heat-resistant suppressors that had simultaneously acquired a cold-sensitive growth phenotype (we will call them Sup/Cs). He also screened Cs morphogenesis mutants for suppressors that had simultaneously acquired a heat-sensitive phenotype (Sup/Ts). This idea had two advantages. First, looking for a new growth phenotype was much more efficient than backcrossing as a means to detect the pseudo-revertants among the revertants; true revertants would not have a new phenotype. When the putative pseudo-revertants were crossed back to wild type, the original mutation reappeared among the progeny, as expected from the genetic definition of suppression. Jarvik discovered that the Sup/Cs or Sup/Ts mutations retained their new growth phenotypes when separated from the original mutations they suppressed. Second, he could use the new growth phenotype, which always turned out to be recessive, in complementation tests with other mutations (usually nonsense mutations that produced null phenotypes) to discover in which functional gene the suppressor mutation had arisen.

This approach had the advantage that it could be iterated. Once separated from the original mutation (let us say it had a Ts phenotype), the suppressor's new growth phenotype (Cs) could be selected against and, once again, suppressors with a new Ts phenotype recovered. Just as before, these new suppressor mutations could be separated from the Cs mutation and its functional gene identified using the new Ts phenotype. Jarvik was able to iterate the process as many as four times.

This approach can be used in other experimental settings; American molecular biologist Don Moir used it later in his studies of genes that turned out to control yeast DNA replication, again with considerable success.[3] Most of the interactions discovered in the phage morphogenesis studies of the 1970s and in the yeast DNA-replication experiments of the 1980s were later verified by direct biochemical analyses.

In the yeast system, it was simple to test dominance and recessiveness of the growth phenotypes of Sup/Ts or Sup/Cs mutants and also of the suppression phenotype. Without exception, the growth phenotypes were recessive to wild

[2] Jarvik J, Botstein D. 1975. Conditional-lethal mutations that suppress genetic defects in morphogenesis by altering structural proteins. *Proc Natl Acad Sci* **72:** 2738–2742.

[3] Moir D, Stewart SE, Osmond BC, Botstein D. 1982. Cold-sensitive cell-division-cycle mutants of yeast: Properties and pseudoreversion studies. *Genetics* **100:** 565–577.

type, whether or not the original mutation was present. In contrast, the suppression phenotypes were dominant.

It is illuminating to consider the heterozygous diploids required to make this assessment. For example, to test a Sup/Cs allele that reverses the phenotype of an original Ts mutation (*tsmut*), the required diploid would be homozygous for the original Ts mutation (*tsmut*) and heterozygous for the Sup/Cs suppressor mutation (*sup/cs*) itself:

$$\frac{tsmut}{tsmut} \quad \frac{sup/cs}{nosup}.$$

The growth phenotype of this strain is not heat-sensitive (Ts), because the suppression phenotype of the *sup/cs* mutation is dominant, and the strain is not cold-sensitive (Cs), because the growth phenotype of the *sup/cs* mutation is recessive to its wild-type allele (*nosup*).

This result is consistent with the assumption that the mechanism of suppression involves interaction, although, as we will see later, interaction suppressors are not necessarily dominant. If we assume that the recessive phenotype of the Ts mutation (*tsmut*) reflects a loss of function of a protein that is restored by its interaction with the mutant protein—a mutant protein that is encoded by the gene in which the *sup/cs* mutation lies—then in the heterozygote, a function-restoring interaction could occur. And if all that is required is a partial restoration of the activity of the affected complex, then function should be restored in the heterozygote.

## Mutual Interaction Suppressors

The morphogenesis, structure, and function of the cytoskeleton of eukaryotic cells, like phage morphogenesis, involve many interacting proteins. Understanding the biology of the cytoskeleton remains one of the central aims of cell biological research. The genetic analysis of functional suppression relationships has contributed substantially to our current understanding of this important cell structure. Indeed, much of what is now known about the many proteins that interact with actin, a key component of the cytoskeleton, was learned from studying mutants in which these proteins were disrupted in yeast.

For some reason, mutations in the gene or genes that encode actin had not been found among the extensive collection of yeast conditional-lethal mutants. However, when the yeast actin gene was cloned in 1980, by taking advantage of the homology between yeast actin and the actins of other eukaryotes, the door was finally opened to genetic analyses of the actin cytoskeleton. Many organisms contain several versions of actin in their genomes (called paralogs; see Chapter 13). However, yeast has only one, simplifying its genetic analysis considerably. Very soon after yeast actin was cloned, it was shown that its deletion was indeed lethal (as everyone had already suspected). Conditional-lethal alleles of actin were then constructed using DNA technology, many of which had a temperature-sensitive (Ts) phenotype.

Many suppressors of Ts actin mutations were isolated in haploid yeast. I will discuss these later because most of them turned out to be recessive, and probably represent losses of function. American geneticist and molecular biologist Alison Adams was determined to find dominant suppressors of actin mutants because she hoped they would be more likely to represent suppression by interaction. To this end, she selected heat-resistant revertants of a diploid strain homozygous for one of these new conditional Ts actin alleles.[4] In such a diploid, potential recessive suppressors would not be recovered. Genetic analysis showed that four of the suppressor mutations she obtained, when separated from the original Ts actin mutation, conferred their own new Ts phenotype. The suppressors were allele-specific, failing to suppress some of the other Ts actin mutations. All of these suppressors mapped to the same locus, and all turned out to be in the gene that encodes fimbrin, which cell biologists had already identified, by biochemical means, as a protein associated with the actin cytoskeleton in their studies of other organisms.

The most remarkable result occurred when Adams selected suppressors of the new Ts phenotype in the fimbrin gene, this time in haploids. She readily found pseudo-revertants (by backcrossing to wild type). Once again, some of these new suppressors were Ts when separated from the fimbrin mutation, and they could be mapped to the actin locus. They were new conditional-lethal alleles of the gene encoding actin! Thus, the suppression is mutual: A heat-sensitive mutation in fimbrin can suppress heat-sensitive mutations in actin, and vice versa. The double mutants are heat-resistant and segregate progeny, upon crossing to wild type, some of which are Ts for actin and some of which are Ts for fimbrin. After much further analysis (including some biochemistry), it is now clear that there is a physical interaction between fimbrin and a particular limited part of the surface of the actin protein structure, where the suppressible (and mutually suppressing) mutations lie.

## Dosage Suppression

This refers to the suppression of mutations in one gene by overexpression of another. Many recessive mutations, particularly conditional-lethal mutations,

<hr/>

[4] Adams AE, Botstein D. 1989. Dominant suppressors of yeast actin mutations that are reciprocally suppressed. *Genetics* **121**: 675–683.

result in the synthesis of mutant proteins that have lost normal activity, stability, and/or affinity for their binding partners and/or ligands. In many of these circumstances, raising the concentration of binding partners can, by the law of mass action, increase the amount of stable active complex and thus restore, to some degree, the function of the mutant protein. Once it became easy to overproduce proteins at will using DNA technology, several investigators found that many mutations, especially conditional (e.g., heat-sensitive or cold-sensitive) mutations, could be suppressed by the overexpression of what turned out to be related proteins.

This kind of suppression has been very effective for detecting highly specific protein interactions. An early, but very persuasive, example was provided in 1991 by the British yeast cell biologist Kim Nasmyth and his group.[5] In yeasts, as in other eukaryotes, the highly conserved MAPK protein (MAPK stands for mitogen-activated protein kinase) is the master regulator of the eukaryotic cell cycle. This protein acts at two points in the cell cycle: at the initiation of DNA replication (S phase) and at mitosis (M phase). By 1991, it was already known that the S phase requires A-type cyclins to bind to and activate MAPK, while the M phase requires B-type cyclins to bind to and activate this protein. The experiment by Nasmyth's group began with a conditional-lethal mutant of MAPK, in which only the initiation of M phase was defective (S phase activation was normal). They then introduced into this mutant numerous plasmids that overexpressed different yeast proteins to identify suppressors of this phenotype. Plasmids overexpressing B-type cyclins were found to suppress the defect in the MAPK mutant, whereas plasmids that overexpress A-type cyclins were not recovered and, when expressly tested, did not suppress this defect.

In the succeeding 20 years, all this has been confirmed biochemically. The mutant MAPK protein has indeed been shown to have a lower binding affinity specifically for the B-type cyclins; the overexpression of B cyclins reverses this defect because the higher concentration of this cyclin compensates for its weaker binding.

More recently, a genome-scale study of dosage suppression of 437 conditional-lethal yeast mutants confirmed that dosage suppression generally involves proteins of related function. It may well turn out that this method, based as it is on the law of mass action, is the simplest and most effective method for finding functionally related genes and proteins.

---

[5] Surana U, Amon A, Dowzer C, McGrew J, Byers B, Nasmyth K. 1993. Destruction of the CDC28/CLB mitotic kinase is not required for the metaphase to anaphase transition in budding yeast. *EMBO J* **12:** 1969–1978.

## Bypass Suppressors

This form of suppression occurs when an essential function is bypassed by providing an alternative gene or protein to perform that function. The classic example of this is a metabolic one. In bacteria, mutations in the proline biosynthetic pathway (including deletions, guaranteed to produce null phenotypes) can be suppressed by mutations that affect enzymes in the biosynthetic pathway of another amino acid, arginine. The pathways that lead to these two amino acids have much in common. An intermediate in proline biosynthesis is a relatively unstable compound called glutamic semialdehyde. A more stable derivative, acetyl-glutamic semialdehyde, is an intermediate in arginine biosynthesis.

The acetylation of glutamate serves to isolate the two pathways from each other so they do not compete for a common unstable intermediate. The acetyl group has to be removed later in order to make arginine and is carried out by an enzyme (E in Fig. 9.1) that can also remove the acetyl group from acetyl-glutamic semialdehyde. However, acetyl-glutamic semialdehyde is not its normal substrate, and this reaction takes place only when there is a large excess of this substrate in the cell.

The mechanism of the bypass is now clear and is illustrated in Figure 9.1: The first and second steps of proline biosynthesis lead to glutamic semialdehyde. If the enzyme that catalyzes either of these steps is missing, the cell becomes a proline auxotroph owing to the lack of this intermediate. An alternative source of glutamic semialdehyde becomes available when the reaction catalyzed by enzyme D in the arginine pathway (see Fig. 9.1) is blocked, even partially, creating an excess of acetyl-glutamic semialdehyde, for reasons explained in Chapter 5. Then, and only then, will the E enzyme have enough substrate to remove the acetyl group and provide the proline pathway with a minimal supply of the missing intermediate.

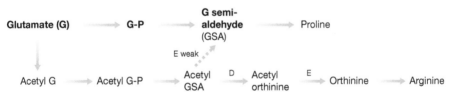

**Figure 9.1.** Proline and arginine biosynthetic pathways. A simplified depiction of the metabolic pathways that create the amino acids proline and arginine. These pathways contain very similar intermediates, glutamic semialdehyde (GSA), which is made from glutamate phosphate (G-P), and acetyl–glutamic semialdehyde (acetyl GSA). GSA and acetyl GSA differ only in the presence of an acetyl group that is removed by enzyme E to make arginine, but it does so inefficiently. Blocking enzyme D results in overproduction of acetyl GSA, which can compensate for defects in the first two steps of the proline pathway.

Thus, a loss of function (even a deletion) in the proline pathway can be suppressed by another loss of function in the arginine pathway. In such a situation, the suppression is likely to be recessive—in a heterozygote in which there is no deficit of the D enzyme, there would be no excess acetyl-glutamic semialdehyde and thus no bypass.

## Recessive Functional Suppressors

Virtually every kind of suppression discussed so far acts dominantly, with the exception of the metabolic bypass of proline auxotrophy. But recessive functional suppression is by no means rare. As discussed earlier in this chapter, suppressors of Ts actin alleles of yeast were often recessive, and we will return to them later.

My favorite illustration of recessive functional suppression concerns the function of the cytoskeleton in the flagella of the photosynthetic unicellular alga *Chlamydomonas reinhardtii*. These organisms swim toward light, and it is easy to find paralyzed mutants that cannot swim. There are many complementation groups of paralyzed mutants, and most of the functional genes they define specify proteins in the tubulin cytoskeleton of the flagellum. This internal structure, which is typical of eukaryotic cilia and flagella, is very complex, consisting of a central pair of microtubules, which are surrounded by nine microtubule doublets and many cross-bridging protein structures that form radial spokes. Mutations that cause paralysis occur in the genes that specify proteins associated with the central pair and in those that specify proteins associated with the radial spokes.

American cell biologist and geneticist Bessie Huang selected revertants of paralyzed mutants that regained the ability to swim toward the light.[6] Four of these turned out to be unlinked suppressors because when each mutant was backcrossed to wild-type individuals, among the progeny were segregants that once again exhibited the paralyzed phenotype. When complementation tests were performed for the suppressor phenotypes, three of them were found to be recessive and the fourth partially dominant. Only one of the suppressors (the partially dominant one) showed a motility phenotype of its own, but in all of them the cytoskeletal structure was altered biochemically. In three of the four cases, one or more proteins found in the normal structure were missing. Clearly, this biochemical phenotype is consistent with the recessive action of the suppressors. It is also

---

[6] Huang B, Ramanis Z, Luck DJ. 1982. Suppressor mutations in *Chlamydomonas* reveal a regulatory mechanism for flagellar function. *Cell* **28**: 115–124.

consistent with the possibility that these suppressors work by bypassing some or all of the function of the central-pair microtubules and the radial spokes.

This idea is strongly reinforced by a remarkable and revealing feature of this system. Two of Huang's suppressors suppress paralysis-causing mutations in many different genes that encode proteins associated with either the central-pair microtubules or the radial spokes. The other two suppressors suppress mutations in many different genes also, but only in those encoding the proteins of the radial spokes.

These results support a kind of bypass model for the action of the central pair and radial spokes in flagellar motion, in which the central pair and radial spokes function to regulate how the flagella beat. The original mutations in these structures cause paralysis because all of the components of the flagellar-motion machinery are connected and a defect in any of them can cause paralysis. The suppressors cause loss of some or all of this connection (and/or of the structures themselves), leaving the surrounding nine microtubule doublets free to beat. This beating may be less efficient than that of the intact cytoskeleton but is sufficient to allow the organism to swim toward the light. Indeed, microscopic studies have shown, in Bessie Huang's memorable phrase, that the intact organism swims a sophisticated, efficient breaststroke, while the suppressed mutants just wave their arms about. As long as the flagellar movement (or lack of it) reacts normally to the sensation of light (through the phototaxis system), some ability to chase the light will be restored.

The functional suppression of paralyzed mutants of *Chlamydomonas* flagella reflects an evolutionary principle that helps to account for the prevalence of recessive functional suppression. The principle is that evolution generally works by improving existing mechanisms, as opposed to replacing one mechanism with another. In this case, one can imagine that the most primitive form of locomotion might have consisted of one or a few microtubules, the movement of which was operated by a motor of some kind. This system would have sufficed to give minimal motility and might have then evolved into a bundle of microtubules, and then into a relatively sophisticated nine-microtubule interconnected molecular machine. This more sophisticated flagellum would have facilitated more efficient motility. Finally, the modern flagellum evolved by the addition of the central pair and the radial spokes, which improved motion into an elegant and efficient breaststroke.

One would expect it to be easy to find mutations in highly evolved and regulated systems that damage regulatory improvements and that also cause the system to fail because all of the system's parts are physically or functionally connected. Such failures can be remedied in two ways: by restoring the damage (a low probability) or by recessive functional bypass suppression, which disconnects or removes entirely the regulatory system, resulting in the partial restoration of function by the more primitive, unimproved system.

The prevalence of recessive suppressors of Ts actin mutations in yeast, mentioned earlier in this chapter, is worth considering in this light. In most cases, when suppressors of Ts mutations with new Cs phenotypes were isolated in yeast genes other than the actin gene, they generally turned out to have dominant suppressor phenotypes. Many were also biochemically verified in time to act as suppressors by affecting protein interactions. The same was true of the dominant mutual suppressors of Ts actin mutations discussed earlier in this chapter. However, the suppressors of Ts actin mutants found by American yeast cell biologist and geneticist Peter Novick in haploids with new Cs phenotypes were all recessive. Actin is a globular protein that consists of large filaments, just like tubulin, which makes up microtubules. It seems quite possible that large filamentous systems such as these provide many opportunities for loss-of-function bypass suppression to occur over time, leaving simple dominant interaction suppressors in the minority.

A final example of recessive functional suppressors is sex determination in the nematode worm *Caenorhabditis elegans*. As I mentioned in Chapter 4, sex determination in the worm is controlled by a signal transduction pathway. The salient property of these pathways is that in double-mutant analysis the phenotype of the mutation that falls later in the pathway masks the phenotype of the other mutation. British geneticist and *C. elegans* developmental biologist Jonathan Hodgkin screened for suppressors of a mutation early in the pathway that results in the masculinization of worms that would otherwise be hermaphrodites, as are wild-type worms.[7] He recovered recessive suppressors in several genes, some of which appear to be null mutations, all of which act downstream of (later than) the gene in which the original mutation lies. The loss of function later in the pathway results in a double mutation and thus has the effect of suppressing the phenotype of the earlier one. Many signal transduction pathways have a similar structure, and thus recessive functional suppression is likely to be found not only in complex molecular machines like the flagella or the cytoskeleton, but also in signal transduction pathways.

To summarize: There are very many possible ways in which mutations can be functionally suppressed. I expect that there will be as many detailed instances of functional suppression as there are types of functional interactions among genes

---

[7] Hodgkin J. 1986. Sex determination in the nematode *C. elegans*: Analysis of *tra-3* suppressors and characterization of *fem* genes. *Genetics* **114**: 15–52.

and their products. Many of the suppressor mechanisms clearly involve direct protein interactions. Cellular systems that involve complicated molecular machinery have provided numerous examples of this kind. However, it is clear from the metabolic-pathway and signal-transduction examples that functional interaction does not necessarily reflect direct interaction among proteins. Either way, genetic suppression presents one way of inferring functional relationships in living cells.

## INTRODUCTORY BIOGRAPHIES

**Costa Georgopoulos (b. 1942)** is a Greek-born American microbiologist and biochemist. He has spent most of his career studying protein folding by the chaperone machines, particularly the one he discovered with Ira Herskowitz. He was a graduate student with S.E. Luria, one of the founders of the field of molecular biology.

**Ira Herskowitz (1946–2003)** was an American geneticist who made transformational contributions to the understanding of the complex regulatory system of phage λ early in his career and to the understanding of the cell biology and regulation of yeast, especially the mating type system. He was a gifted teacher, and he had a lasting influence, directly or through his students, on establishing yeast as the premier eukaryotic experimental system.

**Jonathan W. Jarvik (b. 1945)** is an American geneticist and cell biologist. After working on suppressors in phage P22 for his doctoral thesis, he took up the study of flagellar morphogenesis in *Chlamydomonas*.

**Donald T. Moir (b. 1949)** is an American molecular biologist. After his work on yeast cell cycle mutants, he joined one of the earliest biotechnology companies, where he has worked on human genetics and microbial genetics.

**Alison E.M. Adams (b. 1955)** is a British-born American geneticist and molecular biologist. In her doctoral work with John Pringle, she was the first to visualize the actin and tubulin cytoskeletons of yeast, and thus became a founder of the field of yeast cell biology. She has mainly worked on the genetics and cell biology of the actin cytoskeleton. In recent years, she has taken up sociological studies of clinical trials.

**Kim A. Nasmyth (b. 1952)** is a British yeast cell biologist who has made major contributions to our understanding of the cell cycle in both budding and fission yeasts. He also has made important contributions to our understanding of the regulation of mating type, the role of cohesins in sister-chromosome separation during mitosis, and gene silencing.

**Bessie Huang (b. 1945)** is an American cell biologist and geneticist who spent her entire career on the morphogenesis and function of *Chlamydomonas* flagellae.

**Peter Jay Novick (b. 1954)** is an American yeast cell biologist and geneticist who, in his doctoral work with Randy Schekman, isolated and characterized conditional-lethal mutants defective in protein secretion and defined the yeast secretion pathway, which turned out to be quite general for eukaryotes. Many of the genes in humans retain the yeast names given by Novick and Schekman. After he characterized the phenotypes of the first yeast actin mutations and isolated suppressors of them, he returned to the study of the secretion pathway and pioneered the understanding of the small RAS-related GTPases.

**Jonathan Hodgkin (b. 1949)** is a British geneticist and cell biologist who was one of the early workers on *C. elegans* with Sydney Brenner. He elucidated the signal transduction pathway that functions in sex determination and has remained a major figure in worm genetics and its research community.

# CHAPTER 10

# The Genetics of Complex Phenotypes

Many phenotypes are inherited, or have heritable components. However, there are phenotypes for which no single Mendelian factor emerges that can fully account for their heritability when crosses are done, or, in the case of humans, when families with the phenotype are studied. This is true even when heritability is strongly substantiated by twin and family studies. I have mentioned some possible reasons for this already, such as incomplete penetrance or variable expressivity (see Chapter 2), which can complicate the analysis, as can environmental variables. In this chapter, I will not address these further. Instead, I concentrate here on situations in which a phenotype or trait is caused by more than one gene. As human genetic technology has advanced, it has become clear that very many inherited propensities to human disease fall into this category.

The genetics of complex phenotypes has been both murky and contentious for a long time. It is an area in which many words and ideas have been introduced, many of which have done little to clarify the underlying science, largely because the experimental means have not been available. With the advent of DNA sequencing and genomic analysis, this situation has begun to change much for the better. For this reason, I feel that this is a good time to rethink some of the old words and ideas that prevail in this area of genetics. I have tried to do so here in this chapter, introducing only those words that I think are helpful, and resorting to plain language for the rest.

*Genetic Heterogeneity:* This phrase refers to circumstances in which a particular phenotype can be caused by mutations in any of several genes. For example, there are five genes in which simple recessive mutations will cause tryptophan auxotrophy in yeasts. This is not a problem in model organisms because these different genes can be sorted out by simple complementation analysis, as we have already seen. In human genetics, however, where we cannot do crosses, genetic heterogeneity becomes a problem. In some diseases, such as Fanconi anemia, one family might segregate a causative allele in one locus, while another segregates a causative allele at a second, unlinked locus, and yet the disease phenotype remains the same. When we then use DNA markers to try to map a causal locus, the DNA markers linked to the first locus will not be linked to the second locus (and vice versa). Consequently, the combined LOD statistic used to evaluate linkage will converge on zero instead of rising to significance as it normally would if only one locus was involved.

*Polygenic Inheritance:* When a phenotype is determined by the joint action of two or more genes, its mode of inheritance is called "polygenic." The most obvious phenotypes subject to polygenic inheritance are quantitative traits (see below), although there are many instances in which discrete traits are controlled by gene interactions.

*Quantitative Trait:* These are traits that can vary continuously in a population, depending on the combination of alleles of several different genes that are inherited. Examples of quantitative traits are blood pressure in humans and other animals, and fruit weight in plants. If one crosses a modern tomato (average fruit weight of 100 g or more) with a more primitive member of this species (average fruit weight of just a few grams), one finds in the segregating progeny of the hybrid an essentially continuous, broad distribution of weights, even though the original parents had well-separated distributions of fruit weight. Using DNA polymorphisms as markers, it is possible to parse out individual "quantitative trait loci" (QTLs) that contribute to a quantitative phenotype. For now, it suffices to say that the analysis of quantitative traits by linkage analysis has been effective in highlighting some, if not all, of the genes that contribute to such traits. Clearly, genes that have a large individual contribution are relatively easy to find by QTL mapping, whereas genes with smaller effect are more difficult to identify, even though there may be many more of the latter than of the former. I return to this subject and to QTLs in the context of human genetics in Chapters 14 and 15.

*Synthetic Phenotype:* A synthetic phenotype refers to the novel phenotypic consequences of combining the mutant alleles of different functional genes. The origin of this usage was "synthetic lethality," a term introduced by Theodosius Dobzhansky to describe the phenomenon wherein alleles of two different functional genes, each of which confer a nonlethal phenotype, cause a lethal phenotype when combined together in the same strain of *Drosophila*. Readers will recognize this as a case of two genes contributing to a discrete phenotype, in this case life itself. In recent years, large-scale systematic studies of the relative fitness of double mutants in yeasts and other model organisms have brought synthetic gene interactions to the forefront of molecular genetics.

It will no doubt have occurred to the reader that functional suppression (as discussed in Chapter 9) logically falls under the rubric of synthetic phenotypes. As we will see, functional suppressors are indeed part of the landscape of synthetic phenotypes. To my mind, however, including informational suppression under this rubric would be a real stretch. Informational suppression has an internal logic and mechanistic clarity that separates it from the more general (and often mechanistically murky) world of synthetic phenotypes.

*Epistasis:* In Chapter 4, I briefly introduced this word in the context of pathway analysis. "Epistasis" has been used for a century to refer generally to the phenotypic consequences of interactions between two (or sometimes more) different genes. "Epistasis" is one of those specialized words in genetics that, in my experience, causes more confusion than enlightenment. In the general case, I find "synthetic phenotype" much less confusing to my students than "epistasis."

Since its introduction by William Bateson (see Chapter 1) in 1909 to describe the masking of the phenotype of one mutation by another mutation in a different gene (as explained in Chapter 4), "epistasis" has come to mean very different things to different people. In 1918, Ronald Fisher (see Chapter 3) muddied the water considerably by introducing a very similar word, "epistacy," which soon became conflated with Bateson's word "epistasis." Today, "epistasis" suffers from a zoo of incompatible private usages by experimental geneticists, quantitative geneticists, population geneticists, epidemiologists, and evolutionary biologists. It has acquired modifiers that only increase the confusion: One finds in the literature "positive (or synergistic) epistasis," "negative (or antagonistic) epistasis," "reciprocal sign epistasis," "haploid epistasis," "diploid epistasis," and so on. Rather than try to parse these various factional and unintuitive interpretations of this word, in my teaching (and in this book), I try to avoid the word in any sense more specific than the consequences of gene interactions.

In two limited ways, "epistasis" is still a useful word. First, like most experimental geneticists, I sometimes refer to "tests of epistasis," in which double mutants are constructed to assess the order of gene function in a pathway; I did this in Chapter 4. Even so, I tend to avoid this usage by simply referring to the analysis of double mutants or the masking of phenotypes. Second, quantitative geneticists and systems biologists use "epistasis" to refer to the degree to which double-mutant quantitative phenotypes deviate from expectation, relative to the phenotype of individual, single mutants. I tend to avoid this usage as well by referring instead to synthetic phenotypes. The reader, nevertheless, will encounter all of these usages, and care must be taken not to assume the wrong meaning.

*Gene Interaction:* This phrase is frequently used to refer to the common basis of synthetic phenotypes of all kinds. Unlike "synthetic phenotype," this usage is susceptible to the trivial interpretation that genes or gene products work together to produce a product (e.g., a ribosome or a metabolite), which, while true, can be said of most genes without any reference to the specific genetic consequences of individual mutations or ensembles of mutations. For this reason, I prefer "synthetic phenotype."

The history of the systematic study of synthetic phenotypes begins with suppression (as discussed in more detail in Chapter 9). In the case of mutual interaction suppression of temperature-sensitive (Ts) actin and fimbrin gene mutations in yeasts, the double mutant has a temperature-independent phenotype that is not present in either single mutant. This kind of interaction is referred to as an "ameliorating" interaction because the double mutant has a lesser or weaker phenotype than that expected from either single mutant phenotype.

And this history continues with synthetic lethality. The systematic analysis of synthetic lethality began in yeast, when Peter Novick made double mutants between all the available alleles of the yeast actin gene and the unlinked genes that encoded suppressors of each allele (see also my earlier discussion of this work in Chapter 9). To his surprise, he discovered that some of the resulting double mutants did not survive, even at permissive temperatures.[1] This kind of interaction is today referred to as "exacerbating" or "synergistic" because the double mutant has a more severe phenotype than that expected from either single mutant phenotype. The yeast genetics community quickly realized that one can find mutations in related genes by screening for synthetic lethality, using a mutant with a conditional-lethal defect in that gene. Even more than screens for functional suppression, synthetic-lethal screens have proved to be a rich source of genes and proteins that are related to those already identified and functioning in a biological process of interest.

## Studying Genome-Scale Genetic Interactions in Yeast

It is helpful to think about genetic interactions in a pairwise fashion, comparing the expected phenotype of a double mutant with that of the component single mutants. If one uses a general phenotype likely to be related to overall fitness, such as growth rate, one can construct simple quantitative models that produce such expectations. Any deviation from the expected growth rate by the double

---

[1] Novick P, Osmond BC, Botstein D. 1989. Suppressors of actin mutations. *Genetics* **121**: 659–674.

mutant then becomes the synthetic phenotype. This is a quantification and generalization of suppression (where the double mutant grows better than expected, relative to wild type) or of synthetic lethality (where the double mutant grows more poorly than expected or not at all).

The Canadian yeast geneticists Brenda Andrews and Charlie Boone have used this approach on a genome-wide scale to study pairwise all the viable null (deletion) mutants in the yeast genome.[2] They found a simple multiplicative model to quantitatively estimate gene interactions. In this model, the fractional growth rates of individual mutants relative to the wild type are multiplied to generate the expected growth rate of the double mutant. As an example, if the growth rates of mutants A and B, relative to wild type, are 0.7 and 0.5, respectively, the

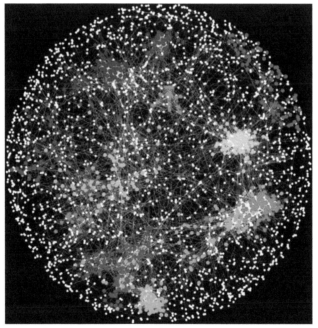

**Figure 10.1.** A network of functional relationships among yeast genes. This network was generated automatically using only the growth phenotype of yeast double-deletion mutants. Each point represents a gene, and the lines between the dots relate to how much the double mutant differs in phenotype from expectation. The colored areas indicate clusters of genes that have been annotated to show they share similar biological functions. (Reproduced from Costanzo M, et al. 2010. *Science 327:* 425–431, with permission, from AAAS.)

---

[2] Costanzo M., Baryshnikova A, Bellay J, Kim Y, Spear ED, Sevier CS, Ding H, Koh JL, Toufighi K, Mostafavi S, et al. 2010. The genetic landscape of a cell. *Science* **327:** 425–431.

expected value for the AB double mutant would be 0.35. With the help of substantial automation, they measured the growth rates of about 10 million double mutants, found many that deviated significantly from expectation, and clustered these together into a network, where the metric for distance between the nodes is related to the magnitude of the deviation (either up or down) from expectation.

Figure 10.1 shows the result of the analysis by Boone and Andrews: a "universe" of genes (white dots) connected by lines (their degree of interaction). Note the colored clusters ("galaxies") of genes that appear close together, which are significant because the analysis of their functions revealed that each cluster is enriched in genes known to be involved in a particular biological process, such as DNA replication, mitosis and chromosome segregation, or translation.

The success of this approach means that the phenotypes of double mutants can be thought of as a new kind of implicit experiment that defines genetic interaction. The network that is created by the ensemble of these experiments is of general interest for all of eukaryotic genetics because most of these genes encode proteins, the structure, function, and interactions of which are conserved through evolution. I will have more to say about evolutionary conservation in later chapters.

## INTRODUCTORY BIOGRAPHIES

**Brenda J. Andrews (b. 1957)** is a Canadian yeast geneticist who made major contributions to the understanding of the regulation of the yeast cell cycle, after which she began a long and productive collaboration with Charlie Boone. Together, they systematically constructed and analyzed the growth phenotypes of ca. 3 million double mutants comprising essentially all pairs of genes in the yeast genome.

**Charlie M. Boone (b. 1960)** is a Canadian yeast geneticist who, with Brenda Andrews, devised and perfected the technical means of making double mutants by recombination on a genomic scale. Unlike with most genome-scale endeavors of this kind, Boone and Andrews found that the patterns of epistasis are biologically interpretable, leading to a new level of functional analysis at the system level.

# Transcriptional Regulation of Gene Expression

In Chapter 7, I introduced the notion that both the activity and the synthesis of proteins must be regulated. From an evolutionary perspective, it seems clear that regulation is required not only to achieve homeostasis (i.e., to maintain a stable internal environment in the face of ever-changing circumstances) but also for efficiency in the use of materials (nutrients) and energy. Regulation mediated by feedback inhibition and by the activation of enzymatic activities in metabolic pathways might be sufficient to achieve metabolic homeostasis over the short term, but changes in protein abundance are required for maintaining homeostasis over longer time periods, even in simple organisms, lest the burden of unnecessary gene expression make cells less fit (in the evolutionary sense) than they could be.

So it is not surprising that regulatory systems that allow cells to change the types and amounts of proteins they contain are a universal feature of all living cells. Such changes in protein content are central to the differentiation of zygotes into multicellular organisms, which contain many different kinds of cells with very diverse protein compositions. The regulation of gene expression is equally important to microorganisms, which must deal not only with short-term metabolic perturbations but also with longer-term changes in their nutritional environment.

Perhaps it is more surprising that the regulation of gene expression—the first step in protein synthesis—is dominated by mechanisms that work at the level of transcription—that is, on the production of mRNA from genomic DNA sequences.

## Instability of mRNA Molecules

A key property of mRNA molecules is their instability; they decay exponentially, usually after having been translated to make protein only a few times. The half-life of a typical bacterial mRNA is a few minutes. The half-life of eukaryotic mRNAs is more variable, but most of them decay more quickly than the generation time of a growing cell. It is this instability that underlies virtually all the mechanisms of transcriptional regulation of gene expression.

For example, the reader may have wondered how the infection of *E. coli* by phage T4 results in the synthesis of phage proteins instead of bacterial ones: This is an extreme case of protein biosynthesis regulation. Following its infection, a living bacterial cell ceases to be able to reproduce itself and is converted into a factory that produces only the T4 virus. This is because shortly upon infection, the host RNA polymerase is inhibited, preventing bacterial DNA from being transcribed. Because the bacterial mRNAs are unstable, just a few minutes postinfection the only mRNA left in the cells is T4 mRNA, and only T4 proteins are then synthesized.

As will be discussed in detail below, the instability of mRNA allows cells to make individual proteins as needed by controlling the transcription of the genes that encode them. When the need for a protein subsides (as it often does in a changing environment), the cell just needs to stop transcribing its gene and any leftover mRNA will decay in less than a single cell generation. This also leaves the protein synthetic machinery free to translate other mRNAs. Proteins, once made, are generally much more stable, and most proteins persist for many cell generations.

There are significant exceptions to these generalizations about the stability of mRNAs and proteins. Nevertheless, in growing cells, stable mRNAs constitute a small minority of all mRNAs, and unstable proteins constitute a small minority of all cellular proteins.

▸ *The relative* abundance *of mRNA molecules generally reflects the* rate *of synthesis of the encoded protein(s) at any particular point in time.*

## Transcription Basics

In all organisms, the flow of information from DNA sequence to protein begins with the synthesis of an mRNA by a complex of proteins that I will refer to collectively as "RNA polymerase." The RNA polymerase complex recognizes sequences in the DNA, binds to them, and transcribes the DNA in one direction. In effect, this means that DNA is transcribed from one of the two DNA strands in the helix, called the template strand (see also Chapter 7), with the mRNA beginning at its 5′ end. The RNA polymerase complex continues along the template strand until it reaches other sequences that cause transcription to terminate, ending the mRNA at its 3′ end.

*Upstream and Downstream:* Thus, transcription moves in the 5′ to 3′ direction, relative to the coding strand. As in the analysis of pathways described in Chapter 4, the flowing water metaphor is in general use when discussing transcription, such that

- "Upstream" refers to DNA sequences in the 5′ direction on the coding strand (see Chapter 7 for an explanation of what the coding strand is). The DNA sequences that bind RNA polymerase generally lie upstream of (i.e., before) the actual point of initiation of transcription. When referring to specific sequence positions, numbering is relative to the first nucleotide in the mRNA, so that −100 means 100 bases upstream of the transcription initiation point and +100 means the 100th nucleotide in the transcript.
- "Downstream" refers to DNA sequences in the 3′ direction on the coding strand.

When bacterial mRNAs are transcribed, the primary transcript is not extensively modified or spliced before being translated. By contrast, most other organisms have a more complicated RNA metabolism that features the formation of a primary transcript, which is then subjected to extensive RNA splicing and other modifications, often in concert with the transport of the mRNA into the cytoplasm, where it can be translated by the ribosomes. Notably, in bacteria, transcription and translation occur together in the cytoplasm, whereas in eukaryotic organisms (including yeasts and humans), transcription occurs in the nucleus and translation in the cytoplasm. Thus, only in bacteria is there coupling between transcription and translation.

Just as the initiation of mRNA translation is determined by the binding of ribosomes to a specific sequence (usually the first ATG codon; see Chapter 7), special DNA sequences, called "promoters," determine where RNA polymerase begins transcription. Gene expression is regulated primarily at the level of transcription initiation, usually by regulatory proteins that bind specific DNA sequences in the vicinity of the promoters.

*Promoter:* According to its original strict usage (as introduced by François Jacob and Jacques Monod in the 1960s; see also Chapter 5), this word refers to the sequences of DNA to which RNA polymerase binds. In *E. coli*, the consensus promoter sequence (i.e., the sequence shared by many *E. coli* promoters) is ... -35 TTGACA ... -10 TATAAT ... on the coding strand (the numbers indicate the number of bases upstream of the beginning of the mRNA). Abundant evidence shows that RNA polymerase binds this sequence and close variants of it. According to this original definition, mutations in a promoter are expected to alter the levels of transcription by either increasing or decreasing the number of mRNAs made per unit time or to abolish transcription entirely. Mutations in promoters as defined in this strict way should not alter the regulation of transcription. This original usage is still common today, for instance, when geneticists refer to "strong" or "weak" promoters.

However, the word "promoter" is also often used today in a looser sense. I think it best to think of the word in this newer sense as a contraction of "promoter region." The promoter region is the stretch of DNA that lies just upstream of (and including) the point at which the transcription of a gene begins. When referring to the promoter region, biologists usually mean the 100 to ∼1000 base pairs immediately upstream of the transcription initiation site. In addition to those sequences recognized specifically by RNA polymerase (i.e., the promoter sequences *sensu stricto*), the promoter region of most genes also contains other DNA sequences that are bound by regulatory proteins that facilitate or inhibit transcription. However, as we shall see, sometimes such DNA sequences can be found elsewhere. I recommend avoiding confusion by using "promoter" according to its original definition, and by using "promoter region" when that is what is being referred to.

## Genetic Analysis of Lactose Utilization in *E. coli*

In my earlier discussion of metabolic regulation at the level of metabolic enzyme synthesis (see Chapter 5), I briefly introduced the fact that *E. coli* cells produce the proteins required to use the sugar lactose as a source of carbon and energy, but only when lactose is present. In 1961, François Jacob and Jacques Monod studied this regulation genetically and biochemically, and in the process laid out the major principles for the regulation of protein biosynthesis at the level of transcription.[1]

As in the case of Seymour Benzer's work on the bacteriophage T4 *rII* system (see Chapter 6), which clarified the nature of the functional gene, the analysis of *E. coli*'s lactose utilization system led to a deep understanding of the principles of transcriptional regulation; these principles apply to all organisms. I therefore discuss the Lac system in some detail in this chapter.

Jacob and Monod invented (or altered) quite a few words that have since become standard usage in biology. Once again, their technical meaning in biology might differ from their usage in ordinary English. They are as follows.

*Inducible:* Inducible refers to those situations in which proteins are only made when a specific chemical compound or metabolite (called the "inducer") is present or in excess.

---

[1] Jacob F, Monod J. 1961. Genetic regulatory mechanisms in the synthesis of proteins. *J Mol Biol* **3**: 318–356.

**Repressible:** By contrast, repressible refers to situations in which proteins are made only when some specific chemical compound or metabolite (called the "corepressor") is absent or limiting.

**Constitutive:** Constitutive refers to situations in which proteins are not specifically regulated, such that protein synthesis is neither inducible nor repressible.

Jacob and Monod, like Benzer, began with the isolation of thousands of mutations. These mutations affect three proteins that are involved in lactose utilization. The synthesis of all three proteins is "inducible," meaning that each protein is synthesized if, and only if, lactose (or a chemical analog) is present in the medium. These three proteins are as follows.

- β-*galactosidase* is the enzyme that carries out the hydrolysis of lactose into glucose and galactose inside bacterial cells. The bacteria can readily use glucose, but they cannot use lactose without this hydrolysis.

- β-*galactoside permease* is a membrane protein that facilitates the entry of lactose into the cells. In its absence, *E. coli* cells cannot grow on lactose, even if they have β-galactosidase in abundance.

- β-*galactoside transacetylase* is another enzyme in bacterial cells that modifies β-galactoside. It is still not clear what this enzyme does for the bacteria, but it is readily assayed biochemically and played an important role in understanding the system because it is coregulated with the permease and with β-galactosidase itself.

Jacob and Monod had at their disposal a number of chemical analogs of lactose (such as chromogenic substrates, which change color when acted upon by β-galactosidase; and "gratuitous" inducers, which cause the induction of β-galactosidase but are not substrates of it). With the help of these analogs, they recovered many mutants with one of two different lactose-specific phenotypes: lactose-negative and constitutive.

**Lactose Negative:** Lactose-negative mutants fail to grow when lactose is the only source of carbon.

**Constitutive Mutants:** Constitutive mutants can grow when lactose is the only source of carbon, but unlike wild-type *E. coli*, they produce all three proteins whether or not lactose (or any of its artificial analog inducers) is present. Jacob and Monod chose the name "constitutive" because in these mutants, the three proteins were regulated in the same manner as are the essential constituents of the cell, such as ribosomal proteins or RNA polymerase. This usage has since become standard.

## Coordinate Expression of the Lac Structural Proteins

A striking result of studying the expression of these three proteins was the observation that they seemed always to be expressed together in the same ratios. In kinetic studies of induction, the three proteins rose in concentration in the cell coordinately. In all the constitutive mutants, the requirement for an inducer to stimulate the expression of all three genes was abolished to the same degree for each. Jacob and Monod were strongly influenced by this observation. Once they realized (from genetic mapping studies) that the three cistrons that encode these three proteins were physically adjacent to each other on the genome, they proposed a simple model in which these three cistrons are transcribed as a single mRNA from a single promoter. Such a transcript is called a "polycistronic mRNA."

*Polycistronic mRNA:* As mentioned above, a polycistronic mRNA is one that encodes more than one cistron. When the synthesis of a polycistronic mRNA is regulated, this regulation automatically results in the expression or inhibition of each of the encoded cistrons, such that the ratio among the proteins is maintained.

*Operon:* An operon refers to a region of the genome in which two or more protein-encoding cistrons are coexpressed and coregulated because they are cotranscribed as a single, polycistronic mRNA. This word, also introduced by Jacob and Monod in their 1961 paper, has become standard in molecular genetics. Operons are very common in all kinds of bacteria, but are rare in eukaryotes. However, not all of the coordinate regulation of transcription occurs as a result of operon structure, even in bacteria.

*Regulon:* An American microbiologist, Werner Maas, introduced the word "regulon" in 1964 to refer to groups of genes that are coregulated by the same regulators but not cotranscribed into a single mRNA. Generally, they are also not near each other in a genome. This usage has become standard and has been expanded to include all kinds of organisms, not just bacteria. A regulon in bacteria might contain several separate genes, as well as some operons. Most higher organisms contain large numbers of regulons but very few true operons. Like most geneticists, I reserve the word "operon" for cases in which a polycistronic mRNA is produced. I discuss the structure of operons in more detail below.

## Genetic Structure of the *lac* Operon

Like Benzer (see Chapter 6), Jacob and Monod made a fine-structure map of their mutations. Also like Benzer, they were fortunate in that all the mutations they studied with lactose-specific phenotypes were closely linked to each other in the same small region of the bacterial chromosome. New methods had also been recently developed that enabled them to make both transient and stable diploids of this region of the bacterial genome, using stable plasmids called "F'*lac*." These methods also allowed Jacob and Monod to carry out complementation, as well as recombination, studies. These included using simple, recessive mutations that could be sorted into four cistrons. They used some of the remaining mutations to define other genetic regulatory elements of the *lac* operon. They placed all of these mutations (and the cistrons and regulatory elements they had defined) relative to each other on the DNA sequence. The recessive, lactose-negative mutants defined three cistrons that Jacob and Monod called the Lac "structural genes":

- *lacZ:* encodes β-galactosidase
- *lacY:* encodes β-galactoside permease
- *lacA:* encodes β-galactoside transacetylase

The recessive constitutive mutants defined the fourth cistron. Jacob and Monod concluded that the constitutive phenotype of these mutants occurs as a result of a loss of function in a gene that encodes a regulator of the operon. The function encoded by this gene must be to prevent the synthesis of the structural genes in the absence of an inducer. For this reason, Jacob and Monod named the product of the *lacI* gene the "Lac repressor."

- *lacI:* a functional gene that encodes the Lac repressor protein

By mapping the remaining mutants that were not simple recessive losses of function, Jacob and Monod also defined two additional regulatory elements that do not encode proteins:

- *lacO:* This is a short genomic region that Jacob and Monod called the "operator." It is the site at which the repressor binds, thereby preventing the transcription of the long, polycistronic mRNA that includes *lacZ, lacY,* and *lacA.*
- *lacP:* This is another short region that Jacob and Monod ultimately called the "promoter," meaning that it is the site at which RNA polymerase binds to produce the single, polycistronic mRNA that includes three structural cistrons: *lacZ, lacY,* and *lacA.*

Today, of course, we have the entire DNA sequence of the *lac* operon and of the many mutants in each element. It consists of a continuous DNA sequence and encompasses several cistrons. These cistrons are bounded by ribosome-binding sites and contain ATG initiation codons at their starts and chain-terminating non-sense codons at their ends (see Fig. 11.1). Most mutations that fall in the *lacZ,*

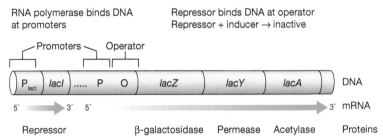

**Figure 11.1.** Genetic structure of the *lac* operon of *E. coli.* The boxes represent the different genetic elements, as also described in the text. The black vertical lines represent cistron boundaries, and the orange arrows show the origin and direction of transcription from the two promoters: the promoter (P) for the Lac repressor (*lacI*) and the promoter for the polycistronic sequence encompassing the three cistrons: *lacZ, lacY,* and *lacA.* The proteins they encode are shown beneath them. When the repressor and inducer bind the operator (O), the transcription of the polycistronic sequence is inhibited.

*lacY,* or *lacA* cistrons cause recessive loss-of-function phenotypes and include missense, nonsense, and frameshift mutations, as well as deletions, some of which cross the cistron boundaries.

A very important feature of the genetics of the *lac* operon is that most frameshift mutants in *lacZ* do not eliminate the function of *lacY* and *lacA,* even though they are transcribed together to produce a single polycistronic mRNA. Thus, the cistron boundaries in the *lac* operon, as in T4 *rII,* are the start and stop translation sequences that are interpreted by the ribosomes. It is worth noting in this context that sequencing studies of T4 mRNA show that *rIIA* and *rIIB* are, like the Lac structural genes, transcribed together into a polycistronic mRNA and thus form an operon.

## Functional Analysis of the *lac* Operon Regulatory Elements

Jacob and Monod found two kinds of constitutive mutants among their *lac* operon mutations. Most mapped to the *lacI* gene and, as mutant haploids, produced the structural proteins β-galactosidase, β-galactoside permease, and β-galactoside transacetylase even in the absence of inducer. They showed that most of these mutations were recessive by constructing partial diploids (using the F'*lac* plasmid) with the genotype

$$lacI^- \ O^+ \ lacZ^+ \ lacY^+ \ lacA^+/lacI^+ \ O^+ \ lacZ^+ \ lacY^+ \ lacA^+.$$

Partial diploids with this genotype produce the structural proteins β-galactosidase, β-galactoside permease, and β-galactoside transacetylase only when induced. *lacI*⁻ mutants consist of the usual mixture of missense, nonsense, frameshift, and deletion mutations that are typical of loss-of-function mutations in ordinary protein-encoding genes. The conclusion of these observations is that the *lacI* gene encodes the Lac repressor.

The phenotype of the remaining constitutive mutations was different. These Jacob and Monod called $O^c$, for operator-constitutive, because they map to a small region now called *lacO* and because, unlike *lacI⁻* mutations, their constitutive phenotype is dominant. Specifically, a partial diploid of genotype

$$lacI^+ \ O^c \ lacZ^+ \ lacY^+ \ lacA^+/lacI^+ \ O^+ \ lacZ^+ \ lacY^+ \ lacA^+$$

requires no inducer to produce the structural proteins β-galactosidase, β-galactoside permease, and β-galactoside transacetylase.

The amount of the Lac structural proteins produced by the $O^c$ mutants depends on the individual mutation. The original Jacob and Monod mutants produced as much as 30% of the fully induced operon, more than enough to display the constitutive phenotype. Just as *lacI* mutations make all three structural proteins constitutive, so do the *lacO^c* mutations. Notably, the ratios of the three proteins produced remain roughly constant. This supports the idea that these mutants, like wild-type bacteria, produce the long, polycistronic mRNA, but at lower levels than the fully induced wild-type cells.

The most remarkable property of the *lacO^c* mutations emerged from a *cis–trans* test. Diploids of the genotype

$$lacI^+ \ O^c \ lacZ^+ \ lacY^+ \ lacA^+/lacI^+ \ O^+ \ lacZ^- \ lacY^+ \ lacA^+$$

(i.e., with the $O^c$ mutation in *cis* to the only functional *lacZ* gene) are constitutive, but *trans* diploids of the genotype

$$lacI^+ \ O^c \ lacZ^- \ lacY^+ \ lacA^+/lacI^+ \ O^+ \ lacZ^+ \ lacY^+ \ lacA^+$$

are not constitutive. Instead, they require inducer in order to express functional β-galactosidase! Additional experiments using antibodies that can detect the inactive mutant β-galactosidase protein showed that this partial diploid constitutively makes the defective β-galactosidase encoded by the defective *lacZ⁻* gene. In other words, the $O^c$ mutation affects only the allele sitting next to it on the DNA; the other allele is still inducible. Similar experiments with mutant *lacY* or *lacA* genes give the same result. This phenomenon is called *cis*-dominance.

**cis-Dominance:** This term refers to a mutation that affects other genes only in the *cis* configuration of a *cis–trans* test. *cis*-Dominance is a general property of DNA sites that are bound by, and respond to, proteins that act on DNA, such as RNA polymerase, and regulators, such as the Lac repressor. This phenomenon is among the most important and far-reaching of the discoveries made by Jacob and Monod. In the *lac* operon, these repressor-binding sites are primarily located upstream of the transcription start site, although there are some binding sites that overlap with the beginning of the mRNA.

Point mutations in the promoter (*lacP*) that alter the binding of RNA polymerase, which is required to initiate transcription, produce the following phenotype: failure to grow on lactose with or without inducer. Like $O^c$ mutations, these mutations are *cis*-dominant; unlike $O^c$ mutations, they affect the strength of the promoter and not its regulation. Thus, partial diploids of the genotype

$$lacI^+ \ lacP^- \ lacZ^+ \ lacY^+ \ lacA^+/lacI^+ \ lacP^+ \ lacZ^- \ lacY^+ \ lacA^+$$

(i.e., with the *lacP*$^-$ mutation in *cis* to the only functional *lacZ* gene) are unable to grow on lactose, but *trans* diploids of the genotype

$$lacI^+ \ lacP^- \ lacZ^- \ lacY^+ \ lacA^+/lacI^+ \ lacP^+ \ lacZ^+ \ lacY^+ \ lacA^+$$

grow as well as wild-type bacteria do.

Most promoter mutants, in the *lac* operon and in other systems, retain a small amount of residual transcription, often just enough to allow investigators to determine that all three of the structural genes in the operon remain coordinately expressed at this low level.

DNA sequence analysis of mutants and binding studies using purified proteins show that, in the *lac* operon, the DNA-binding sites for RNA polymerase (*lacP*) and for the repressor (*lacO*) partially overlap, although the main body of the operator lies downstream of the promoter. This feature strongly supports the general idea that the Lac repressor works, at least in part, by physically occluding the promoter so that RNA polymerase has difficulty binding and/or initiating transcription in the presence of bound repressor. This is a general property of transcriptional repressors.

## Deletions That Remove *lacI*, *lacO*, and *lacP*

Among the mutations that Jacob and Monod recovered were some very extensive deletions, which they mapped by recombination. For modern geneticists, among the most interesting are those that have one end just to the right of *lacO* (removing the promoter but only a very small part of *lacZ*, such that functional β-galactosidase is still made) and extend leftward, removing all of *lacO*, *lacP*, and *lacI*, and beyond. The phenotype of these deletions is striking. They express all three *lac* operon proteins constitutively (as expected, because the repressor and operator were deleted). Surprisingly, however, they also become purine auxotrophs!

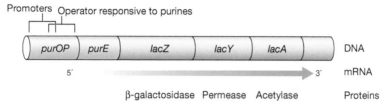

**Figure 11.2.** Genetic structure of *lac* operon deletion mutants. A deletion encompassing the intervening sequence fuses the *purE* regulatory sequences to the Lac structural genes, placing them under new transcriptional control, as shown by the orange arrow.

This is because these deletions remove part of a gene in another operon, called *purE*. The two cistrons in this operon encode proteins that together catalyze a step in the biosynthesis of the purines adenine and guanine, which the reader will recall from Chapter 7 are bases found in DNA and RNA. Furthermore, the amount of the Lac structural proteins that these two types of deletion produce depends on the level of purines in the medium: The expression of the three Lac structural genes is high when purines are limiting and low when purines are in adequate supply. Once again, the three Lac structural genes are expressed together in the usual constant ratio, indicating that they are still expressed in these deletion mutants together as a polycistronic mRNA.

These deletion mutants were arguably the best contemporary evidence for Jacob and Monod's operon theory. As shown in Figure 11.2, the deletions that removed the *lacI* gene fused the Lac structural genes to the operator and promoter of a different operon—one containing *purE*, which forms part of the purine regulon and is controlled by the level of purines in the cell. The normal regulation of the *lac* operon (induction by lactose) is replaced in this deletion mutant by the regulation of the purine pathway (repression by the purine end products). This deletion introduced the idea that the genomic sites responsible for transcription initiation and regulatory action are separable from the cistrons to be transcribed.

## Positive and Negative Control

In principle, there are two ways in which transcription might be controlled by a regulatory protein. In the case of the *lac* operon, as we have seen, when there is no lactose in the medium, the Lac repressor binds to the operator and inhibits transcription. The basic molecular function of the repressor is thus a negative one. Upon the addition of lactose or of another inducer, the repressor undergoes an allosteric change in conformation (see Chapter 5) and can no longer bind the operator. Transcription is now possible because the repressor can no longer carry out its function.

**Negative Control:** This refers to instances in which a regulator, when bound to its DNA sites, prevents or inhibits the production of an mRNA.

**Repressors:** These are molecules that, when bound to their sites on DNA, inhibit transcription that would otherwise occur in their absence.

**Corepressors:** Corepressors are small molecules that, in some negative-regulatory systems, are required to maintain repressors in their active conformation. For instance, the pathway that leads to the biosynthesis of purines is a multioperon regulon under negative control by a repressor. When bound to its corepressors (the purines hypoxanthine and guanine), the purine repressor is active, binds DNA at the several purine regulon operators that are scattered around the genome, and prevents transcription at each of them. When the cells have inadequate supplies of purines to saturate the purine repressor's binding sites, an allosteric conformation change occurs and the repressor becomes inactive.

**Inducers:** In negative-control systems, inducers are small molecules that bind to repressors, rendering them inactive, usually by preventing them from binding DNA. The inducers of the *lac* operon work in this way, as we have already seen.

The alternative mechanism of regulation, positive control, is exemplified in *E. coli* by the arabinose (*ara*) operon. The *ara* operon, like the *lac* operon, is expressed as a polycistronic mRNA, comprising three cistrons that encode proteins required to metabolize arabinose. These cistrons are only expressed when arabinose (or another inducer) is present. In the early 1960s, shortly after the first publication of the great *lac* operon papers, the American microbiologist Ellis Englesberg found evidence that the regulator of the *ara* operon is not a repressor, like the Lac repressor, but functions instead as an activator of transcription. This regulatory protein (the product of the *araC* gene) is required for the transcription of the operon. It only binds its operators and becomes functional to activate transcription after an allosteric change has occurred in the conformation of the activator protein when it binds to arabinose. When the inducer is absent, RNA polymerase does not recognize or transcribe the *ara* operon.

As one would expect, recessive (loss-of-function) mutations in *araC* result in an arabinose-negative phenotype and not in a constitutive phenotype, because loss of the protein results in the inability to transcribe the operon. Constitutive mutants can be found at the sites where the AraC protein binds, but these are *cis*-dominant. This is an example of what is now generally called "positive control" by an "activator," because mechanistically, regulation is accomplished by activating transcription.

**Positive Control:** When a regulator, once bound to its DNA sites, activates or otherwise stimulates the transcription of mRNA, the effect is called positive control.

**Inducers:** In positive-control systems, inducers are small molecules that allow the regulator (the activator) to adopt and maintain its active conformation.

It is worth noting that there are alternative systems of positive control that work not by controlling transcription initiation but rather by controlling its continuation past a particular point in the DNA. For example, bacteriophage λ, which grows on *E. coli*, positively controls several operons by a mechanism called "antitermination." In the absence of the positive-control regulator, transcription terminates before the full polycistronic mRNA is complete (usually only a very short transcript is made); in its presence, termination is aborted and transcription continues.

I noted above that fusion of the *ara* operon to the Lac promoter/operator region results in the control of the arabinose operon by the Lac repressor, and not by the *araC* protein activator. The reader should note that in this situation, the arabinose operon comes under negative control, just like the intact Lac operator in wild-type cells. "Positive" and "negative" refer only to the mechanism of transcriptional control by the regulator, and not to the overall regulatory effect. In wild-type *E. coli*, both the lactose and arabinose operons are *inducible* systems that avoid the synthesis of structural genes when there is no lactose or arabinose for them to consume. In contrast, the biosynthesis of purines, which is also under negative control, is a *repressible* system that avoids the excess synthesis of the purine biosynthetic enzymes under normal conditions, but results in increased expression when purines (the corepressors) are limiting.

## Regulation by Multiple Inputs

The expression of many genes is regulated by more than one system of regulation. Even the simplest metabolic systems have evolved to respond to more than one input in order to maintain homeostasis and balanced growth and thus optimize their relative fitness over evolutionary time. Many genes in bacteria, and virtually all genes in higher organisms, are regulated by two, and sometimes many more, regulators of transcription.

For example, not only are the *lac* and *ara* operons controlled by the Lac repressor and the Ara activator, respectively, which act in *trans* on their specific binding sites, but both are also regulated by another protein, called the "catabolite activator protein" (CAP). CAP is a *trans*-acting, positive regulator of transcription that binds to specific sites (short DNA sequences different from the Lac or Ara operators) that are present in many different promoter regions. Once bound, CAP then stimulates transcription. The *cis*-acting, CAP-binding sites are present in promoters that control proteins, like Lac and Ara, that allow an organism to use alternative sources of carbon and energy. A small-molecule ligand (cyclic AMP, also called cAMP) is required to maintain CAP in its active conformation, just as arabinose is required to maintain the AraC protein in its active state.

There is a physiological advantage to the cell of having this second regulatory system. cAMP is abundant in those cells that are growing on a relatively poor carbon source or are starving for carbon. When cells are growing on a preferred carbon source (such as glucose), cAMP levels are low. So CAP activation stimulates the transcription of inducible systems that utilize less-preferred sugars only when cells have limited access to carbon and energy. It is kept inactive by low cAMP when carbon and energy are abundant. Both lactose and arabinose are poorer carbon sources than glucose, so this second regulatory system serves to stimulate the transcription of the *lac* and *ara* operons only when the cells have no better carbon source available to them.

This combined system, consisting of the *trans*-acting, operon-specific regulators that work at the *cis*-acting Lac and Ara operators, plus the additional, *trans*-acting CAP activator, which works at *cis*-acting binding sites present near the promoters of both, guarantees that the operons will be induced only when two conditions are fulfilled: an alternative nutrient (lactose or arabinose) is available, and a better nutrient (glucose) is not. This is an elegant, logical circuit that makes physiological and evolutionary sense.

**Combinatorial Regulation:** This phrase refers to the transcriptional regulation of the same gene or genes by two or more regulators simultaneously. It entails the presence of two or more regulatory proteins and their *cis*-acting binding sites. Combinations can involve both positive and negative regulators.

As one might well imagine, there are, especially in higher organisms, promoter regions that need to be regulated by more than two input signals. Indeed, genetic and physiological studies have shown that this is the general case for most eukaryotic genes. Some are regulated by up to a score of *trans*-acting regulatory proteins that recognize *cis*-acting sites on the DNA. However, these sites are by no means all located in the immediate vicinity of the promoters of these genes, as is found in the *lac* operon. Instead, various mechanisms have evolved to enable *cis*-regulatory sites to be located some distance away from the promoters they control. These mechanisms usually involve the bending (or looping) of DNA, which serves to bring together the various regulators of a combinatorially regulated gene so that they can exert their joint influence on the RNA polymerase.

For historical reasons, even though they act in *cis*, these distant regulatory DNA sequences are called not "operators," but "enhancers" (if the effect of protein binding is to stimulate transcription) or "silencers" (if the effect of protein binding is to inhibit transcription). Enhancers and silencers can be quite far upstream or downstream of the ORFs the expression of which they control, but rarely are they more than a few genes distant. They are also often located in introns of the regulated gene or of one of its immediate neighbors.

The reader should be able to infer, at this point, the genetic properties of enhancer and silencer mutations: Being sites of protein binding, such mutations are generally *cis*-dominant. Mutations in the genes encoding the cognate regulatory proteins are generally simple recessives.

Given the complications of combinatorial control by multiple factors, it is not surprising that the functional RNA polymerase complexes of eukaryotes can be very elaborate, containing dozens of subunits, many of which are "transcription factors" that bind specific enhancers or silencers. It is also not a complete surprise that many of the factors and enhancers that they bind do not have nearly as dramatic effects on transcription as their Lac or Ara counterparts, although some clearly do. Over evolutionary time, there must have been many layers of incremental progress, such that a given factor today, although evolutionarily important, may display relatively little by way of a mutant phenotype by itself.

This summary of transcriptional regulation should not be taken as more than a basic introduction, even concerning the genetics. For instance, I did not introduce specifically the effects of mutations in *trans*-acting regulators that have lost either affinity for DNA or affinity for their small-molecule ligands (inducers or corepressors). It suffices to note here that mutants in the Lac repressor that no longer recognize inducer are, as expected, not inducible under ordinary conditions, and mutants in the Lac repressor that no longer recognize their DNA-binding sites (operators) are constitutive. In contrast, mutants of the AraC activator protein that do not recognize arabinose cannot grow on arabinose, as expected, and mutants in this protein that do not recognize their DNA-binding sites also cannot grow on arabinose. Historically, the *araC* cistron was first defined on the basis of complementation tests among many recessive, arabinose-negative mutants. Further analysis revealed that the *araA*, *araB*, and *araD* cistrons encode the proteins essential to the metabolism of arabinose (they are "structural" genes, in the language of Jacob and Monod), whereas *araC* encodes the activator protein. The primary phenotype, failure to use arabinose as a source of carbon and energy, is common to loss-of-function mutants in all four cistrons.

I also avoided venturing into the realm of dominant-negative mutations in the regulators. These are fascinating and were important historically in coming to an understanding of each of the systems that have been studied in detail. They are, however, a distraction to my purpose here, because they increasingly depend on the unique molecular biology details of each system. So, with some regret, I leave it to the reader to discover more about these mutations elsewhere.

## INTRODUCTORY BIOGRAPHIES

**Werner K. Maas (b. 1921)** is an American microbiologist who was a major contributor to the study of metabolic feedback regulation in bacteria. He made a lifetime's study of the arginine biosynthetic pathway in *E. coli* and its regulation.

**Ellis Englesberg (1921–2013)** was an American microbiologist who discovered that the enzymes responsible for arabinose degradation in *E. coli* are under positive control, in contrast to the lactose-degrading enzymes. He encountered very strong skepticism because of the great impact of Jacob and Monod's success with the Lac system. Englesberg eventually prevailed, using the very same genetic methods introduced by Jacob and Monod to arrive at different results.

# CHAPTER 12

# The Modular Architecture of Genes and Genomes

The essential elements of regulated gene expression, as exemplified by the Lac and Ara systems of *E. coli*, reflect general and universal principles concerning the architecture of genomes. In this chapter, I review these principles and discuss their implications for two apparently disparate areas of genetic research: genetic engineering technology and molecular evolution, particularly relating to the origin of genetic diseases.

## Initiation, Elongation, and Specificity in Macromolecular Synthesis

Complete genome sequences have now been determined for organisms drawn from every branch of the tree of life, and their analyses confirm that in all of these diverse organisms, genomes are transcribed and translated by essentially identical cellular machinery. The evolution of the replication, transcription, and translation machinery has been very slow, reflecting that this machinery is under strong functional constraint (as no organism can dispense with any of its functions). It is therefore not surprising that this machinery, consisting of proteins and RNAs, is indeed basically the same in all organisms.

Some components of this machinery are not specific to any particular gene or locus; I will refer to these as "generic" components to distinguish them from those that recognize particular sequences, thereby providing specificity for one or more sites in a genome. Examples of generic components are the catalytic subunits of RNA polymerases, which automatically copy a strand of DNA once transcription has been initiated at a promoter; the ribosomes, which automatically assemble polypeptides under the direction of the mRNAs, starting generally (but not always) from the first AUG codon; and the catalytic subunits of the DNA replication machinery, which automatically attach to a DNA polymerase once it has started replicating DNA.

However, this cellular machinery does not begin working at random. Its generic components function together with its "specific" (or "regulatory") components (usually proteins and/or RNAs), which guide the complete machine, in the form of a functional complex, to particular sequences on the DNA or RNA where transcription, translation, or replication should occur. In this way, genomic DNA sequences encode not only the content of biological macromolecules but also the detailed circumstances under which each one is made.

Biochemists and molecular geneticists have developed specific words to help separate the sequence-independent and sequence-specific activities that are involved in the biosynthesis of the major biological macromolecules, as follows.

*Initiation:* This term refers to the process by which the synthesis of a macromolecule begins. In intact cells, the initiation of DNA, RNA, or protein synthesis begins only at particular sequences that are recognized by the specific components of the machinery. The sequences that signal the start of transcription are called promoters, as we saw in Chapter 11. The initiation of translation is usually at the first ATG codon in an mRNA, as we saw in Chapter 7. The specific components of the machinery recognize the sequences in the DNA or RNA and activate the machinery to begin polymerization, thereby conferring sequence specificity to the biosynthetic process. Their activity is modulated by other factors that bind DNA in the vicinity—these include activators, repressors, enhancers, and sequence-specific binding proteins (or RNAs) of many other types.

*Elongation:* Elongation is the process of polymerization that ensues once synthesis has been initiated. In general, elongation is not sequence-specific, and continues until a termination signal or sequence is encountered. For protein synthesis, this signal is one of the three "nonsense" or "termination" codons: UAA, UAG, or UGA (see also Fig. 7.3). In most organisms, there are also proteins (termination factors) that recognize stopped ribosomes and release the completed polypeptide. The termination signals for transcription and DNA replication are more complicated, and the termination proteins more essential, but the idea is the same: Elongation automatically continues until specific sequences are encountered that signal termination.

> ▶ *Specificity in macromolecular synthesis is exerted at initiation, at termination, and also (sometimes) during RNA splicing, but not in elongation.*

The DNA-binding proteins that allow transcription to be initiated bind to specific, generally very short (~10- to 20-bp) DNA sequences that mark the sites at which action is required. It is these DNA sequences, and the proteins or RNA molecules that bind to them, and only them, that give transcriptional regulation its remarkable specificity. This is how an essentially promiscuous RNA polymerase can be harnessed to transcribe the *lac*, *ara*, and *purE* operons of *E. coli* independently, as explained in Chapter 11. Each transcription initiation site is regulated by a unique set of repressors and/or activators. It is these specific binding sites that account for the properties of the independent action of the promoters and operators of the *lac*, *ara*, and purine operons.

Most organisms contain some regulators that bind to only one site in the genome and other regulators that bind to many sites. One site per genome means one site in 4 million bp for *E. coli* and one site in 3.3 billion bp in the human genome. How the specific components of the machinery find and bind to such sites quickly and efficiently is still not completely understood, and continues to pose a major question in molecular biology.

One way to think about this system is that these local, short DNA sequences are "addresses" in the genome and the regulators (such as the repressors and activators) are "readers" that scan the DNA for their specific addresses and bind to them when they are found. Some addresses (like the ones recognized by the Lac repressor or the Ara activator) are unique. Others are associated with a few different promoters, such as the purine repressor-binding sequences found in the "regulon" (the operons and genes) that encode the proteins of *E. coli*'s purine biosynthesis pathway. The presence of a binding site for the purine repressor renders the transcription of each member of the regulon sensitive to repression when purines (the corepressors) are in abundance. Yet other addresses are more frequent, like the dozens of catabolite activator protein (CAP)-binding sequences that are present at or near the promoters of genes, which encode proteins involved in using less-preferred carbon sources when glucose is unavailable. (You may recall from Chapter 11 that CAP is a *trans*-acting, positive regulator of transcription that binds to specific sites present at many different promoter regions.)

The address analogy also works well for thinking about combinatorial control. This is where a single gene is regulated by two or more regulators simultaneously and independently. When several addresses are present at a gene's promoter region and are in *cis* to one another (as generally occurs in eukaryotes), the promoter region will be bound by a combination of regulators. As a consequence, transcription proceeds only if the right combination of regulators is present and the level of initiation can be modulated by the ensemble of bound proteins. For example, in Chapter 11, we saw that *lac* operon transcription is maximal if, and only if, the Lac repressor is absent and CAP is present. Both have addresses in the operator/promoter region of the operon.

It is also worth noting here another useful analogy to the logic that underlies computational circuits. In this analogy, if binding of either of two sites by a regulator suffices to cause transcription, this is the equivalent of the logical operator "or." If two sites must both be bound for transcription to proceed, it corresponds to the logical operator "and." Much of transcriptional regulation can be modeled along these lines. Many in the emerging field of "systems biology" have extended the analogies between biological and electronic circuits, producing illuminating and useful insights into principles that underlie biological phenomena at the systems level. Any discussion of "engineering design principles" in biology must, of course, be tempered by the fact that all organisms are the result of evolution; such principles are not driven by any design but instead by selection of the fittest variants among a competing population of organisms.

## Separable Regulatory Sites and Coding Sequences

For me, the most remarkable feature of the regulatory architecture of genomes is their modular regulatory structure. Gene-specific binding sites (the "addresses") are almost always positioned in such a way that they, and the information they contain, are physically separable from the DNA sequences that they regulate. This makes it possible for the structural genes of the *lac* operon to come under new regulation just by replacing the Lac operator/promoter with its counterparts from the *ara* operon or from the *purE* operon (see Chapter 11 for more detail).

Conversely, virtually any DNA sequence, even one that is completely nonnative or synthetic, can be transcribed in *E. coli* if fused to the Lac operator/promoter region. Just as with ribosomes, which will translate any sequence preceded by an in-frame initiator (ATG) codon, RNA polymerase will transcribe any sequence preceded by a functional *E. coli* promoter region. If that region also contains sites recognized by regulatory proteins (such as activators or repressors), then transcription will occur under their regulation. Another way of saying this is that most gene-encoding DNA sequences are physically separated from the *cis*-acting promoter, operator, enhancer, and/or silencer sequences that are recognized by their regulators.

> ▶ *Functional sequences are arranged in genomes essentially as inter-changeable segments. In this way, virtually any sequence can be brought under any type of regulation by placing it in proximity to the appropriate regulatory cis-acting binding sites.*

By its nature, the complete multisubunit transcription machinery (sometimes called the "RNA polymerase II holoenzyme" in eukaryotes) will function at any sites it recognizes and binds to. Thus, in a diploid cell with two copies of a particular set of specific initiation sites, both sites will be acted on. These sites of action, however, only affect the sequences physically attached to them on the DNA; that is, they function only in *cis*. This explains the *cis*-dominant phenotypes produced by mutations in promoters, operators, enhancers, silencers, and origins of replication: These mutations affect only sequences on the same molecule of DNA on which they reside.

It is a good general rule that cellular machinery (made of proteins and RNAs that can diffuse around the cell) acts in *trans*, whereas sites of specificity can act only in *cis*. Consequently, *cis*-acting mutations virtually always turn out to be in sites of action. For transcription, such sites are the promoters that bind RNA polymerases, the operators and enhancers that bind activators or repressors of transcription, and the termination sequences that cause RNA polymerase to stop transcribing. For DNA replication, such sites of action are called "origins" and "terminators" of DNA replication. It should be clear that for these activities to be specific to, for example, only a single locus like the *lac* operon, there must be a high degree of discrimination in DNA binding: The functional elements of the machinery have to find and bind specifically and uniquely at the correct, short sequences in the DNA.

> ▶ *Mutations with cis–dominant phenotypes generally occur in sites that must be recognized specifically so that catalytic proteins, like RNA polymerase, can act nearby on the same molecule without additional sequence specificity.*

## Genomic Architecture and the Rise of Biotechnology

The great advances in molecular biology, such as those that fall under the headings of "genetic engineering" and "recombinant DNA technology," took place immediately after it became clear that macromolecular synthesis is performed nonspecifically by generic machinery under the direction of regulatory molecules that guide the synthetic apparatus to specific, short sequences of DNA.

*Human Protein Production in* **E. coli***:* The discovery of the stepwise manner in which macromolecules are made (as discussed in the preceding section) hinted at the possibility of producing rare and valuable protein molecules in simple organisms, like bacteria or yeasts. Some proteins were already known to be commercially valuable, such as human insulin; at the time, bovine and porcine insulin were used to treat diabetes. Other proteins were so difficult to purify that they became impossibly expensive: A few micrograms of interferon, for example, for research use cost the best part of a million dollars to purify from natural sources. As sequence information became available (either from proteins or from cloned coding sequences), many researchers realized that it might be possible to express these proteins in bacteria or yeasts, simply by joining together a DNA sequence encoding the desired protein to other sequences that direct replication, transcription, and translation in a convenient host such as *E. coli*.

To this end, only a few things were required, all of which quickly became available in the 1970s:

- A fragment of DNA containing a nucleotide sequence that encodes the amino acid sequence of the desired protein. The most convenient way to obtain such fragments today is by direct chemical synthesis.

- A "vector," which is a circular DNA plasmid of bacterial origin that can replicate in *E. coli*. When any other DNA (such as a DNA fragment encoding a desired human protein) is inserted into this vector, the DNA replication machinery of its bacterial host will initiate replication, by recognizing the plasmid's initiating signal for DNA replication (called "the origin of DNA replication"). Because elongation is nonspecific, the insertion of a DNA fragment that contains a human protein–encoding sequence in *cis* will result in its replication along with the rest of the plasmid's DNA. Plasmid vectors also contain one or more bacterial genes (typically ones that confer resistance to an antibiotic) that can be used to select *E. coli* cells that contain the plasmid, thereby also selecting passively for the human protein–encoding DNA joined to it.

- A strong promoter, one that works in *E. coli*, placed upstream of the human protein–encoding sequence in the plasmid, so that it will be transcribed into an mRNA. A bona fide transcription termination site is also helpful, but not absolutely necessary. Early experiments used the Lac promoter, which had an additional advantage: If one included the operator sequence, one could control transcription with inducers of the *lac* operon. This was useful in situations in which the foreign protein being produced might be toxic to *E. coli*.

- A ribosome-binding site and initiator codon (ATG) in frame with the sequence encoding the desired protein, and a termination codon (TAA, TAG, or TGA) in frame at the end, so that translation will initiate and terminate in the right places, producing just the desired polypeptide.

This scheme simply exploits the basic principles of molecular biology and genomic architecture I have discussed above. Nevertheless, it sufficed to make human insulin and human interferons, thereby kick-starting the multibillion-dollar biotechnology industry. Early genetic engineers used the natural *cis*-acting recognition information from *E. coli* (the origin of plasmid DNA replication, a strong promoter, translation initiation, and termination codons) that was required to guide the bacterial machinery to the inserted human DNA sequence and then replicate, transcribe, and translate it, as if it were just another bacterial gene.

The source of the sequences encoding the proteins to be produced has changed over time. The first experiments in recombinant DNA technology involved making DNA copies of mRNAs from natural sources, a procedure that engendered a great deal of opposition, in the United States and elsewhere. A major concern was the possibility that the procedure might be dangerous and that harmful sequences, such as copies of cancer-causing viruses, might contaminate the desired ones. This concern was particularly acute with respect to pharmaceutical uses of the technology. To allay this concern, insulin was first made by producing the required DNA sequences synthetically, using no material (only information) from natural sources. This somewhat ameliorated the political concerns surrounding recombinant DNA technology and no doubt hastened the approval of human insulin as a drug. Today, synthetic DNA is commonly used for pharmaceutical purposes, not so much for social and political reasons, which have abated considerably (although there remains opposition to food derived from genetically modified organisms), but because the cost of producing synthetic DNA has fallen dramatically as technology has improved. There is probably no better indication that biology has become an information science than the fact that today one can produce any protein provided its amino acid sequence is known.

Recombinant DNA technology has, of course, become much more sophisticated and flexible over time. Today, hundreds of proteins are produced on an industrial scale in a variety of hosts besides *E. coli*: Notable among these are animal cells in culture, insect cells in culture, plants (e.g., tobacco), yeast, and bacteria. Today, it is often the case that the proteins being produced are designed variants rather than those found in nature; this is especially true of therapeutic antibodies and their derivatives. The success of this approach in so many diverse situations illustrates forcefully that the ability to separate the coding sequence from the transcription and regulatory apparatus is a very general property of protein-encoding genes in virtually all species.

**Promoter Fusions:** The realization that virtually any protein can be expressed from any regulatory sequence in the same or a different species led to another important advance: the technique of fusing promoter regions to marker proteins to study gene expression in intact cells and animals. The most popular marker proteins are those that are easily visualized experimentally, such as β-galactosidase from *E. coli* or green fluorescent protein (GFP) from the jellyfish *Aequorea victoria*. Methods have been devised to target the coding sequence of a given

**Figure 12.1.** The tissue-specific expression patterns of marker genes. (*Left*) A litter of young mice, three of which express green fluorescent protein (GFP) in their skin cells. (*Right*) A *Drosophila* embryo that expresses GFP under the control of a developmentally regulated promoter and enhancers, which restrict GFP expression to alternative segments in the developing embryo. (*Left*, Reprinted, with permission, from Kubota H, et al. 2004. *Proc Natl Acad Sci 101:* 16489–16494, © National Academy of Sciences, U.S.A. *Right*, Adapted from Wang X, et al. 2010. *PLoS ONE 5:* E11498.)

marker protein into virtually any gene of any organism. When the coding sequence used is complete and includes the ATG codon that initiates translation, one can see where and when the genes controlled by the regulatory promoter sequences are expressed. To date, thousands of different promoters in hundreds of species have been studied in this way. Figure 12.1 shows two examples: mice expressing GFP in their skin cells, driven by a skin-specific set of promoters and enhancers; and a *Drosophila* embryo expressing GFP in a spatially restricted pattern that shows the origin of the segmental body plan of flies, an iconic image in the field of developmental biology.

***Protein Fusions:*** It is also possible to insert coding sequences into experimental organisms that lack the first ATG. When the inserted sequence encodes a marker protein, it will only be expressed if it fuses to a host protein. These kinds of protein fusions have many uses. Among them is the visualization of the subcellular localization of proteins: For example, fusions to membrane proteins result in the visualization of the membranes. Another important use is to produce pharmaceuticals with the properties of two proteins. A prominent example of this is etanercept, a fusion between a receptor protein and an antibody, which has become a major drug in the treatment of autoimmune diseases.

## The Evolutionary Consequences of Genome Architecture

The process of evolution depends on genetic variation—that is, the differences in genomic DNA sequence that exist among individuals in a population. Natural selection favors individuals with genotypes that provide higher reproductive fitness; as a result of this selection, these fitter genotypes become more frequent in the population over time. Although a full discussion of evolutionary genetics is beyond the scope of this book, I want to summarize a few notable features that relate directly to the architecture of genomes and the way in which functional sequence elements can be recognized in them.

The modular architecture of genes and genomes, in which protein-coding sequences are separated from their regulatory sites, makes possible a kind of mutation in which the regulation of protein expression is changed, like the *purE–lacZ* fusions discussed in Chapter 11. The adjacency of regulatory sites and coding regions also allows genomic rearrangements to occur without changes in the expression of more than a few genes. As I noted in Chapter 7, these kinds of mutations are prevalent in populations, occurring at frequencies even higher than point mutations. So it should be expected that mutations that change the pattern of expression of proteins might be as important in providing variation on which selection can act as changes in the proteins themselves.

The elucidation of the complete genome sequences of numerous organisms from across the tree of life provides evolutionary studies with a wonderful resource. However, inferring exactly which types of mutations have provided fitness advantage over evolutionary time is a difficult and uncertain task. One way of gaining insight is to study evolutionary change in situations of ongoing sequence variation and selection over relatively short periods of time. Recent research has provided two biological situations in which sequence changes that are occurring under selection can be assayed and examined. One assay involves continuously growing a microorganism, such as a bacterium or yeast, in the same medium in the laboratory. These organisms grow so quickly that it is feasible to follow their evolution over thousands of generations under relatively constant and reproducible selection conditions. Another approach is to assay the genomic sequences of tumor cells, which arise as a result of an evolutionary process in which a single, aberrant cell evolves into a malignant tumor. This also occurs relatively rapidly, probably over about 1000 or more cell generations, and under relatively constant conditions.

The results of both of these assays are remarkably similar. One finds, both in tumors and diploid yeasts grown in the laboratory, the occurrence of numerous point mutations, relatively few of which appear to confer much by way of improved reproductive fitness. The major recurring mutations that confer big fitness advantages are chromosomal rearrangements, notably copy number variations (particularly deletions and amplifications) and translocations (both reciprocal and nonreciprocal), as well the gain and loss of entire chromosomes (see Chapter 7 for more on these chromosomal abnormalities).

A simple interpretation of these results is that the selective forces that act on evolving cells in these circumstances select more for changes that alter the abundance of individual gene products (protein or RNA) rather than their quality. For example, if a yeast strain in the laboratory selection environment is growth-limited by a nutrient, the simplest solution available by mutation is to increase the ability to take up that nutrient by producing more of the relevant transporter per cell, rather than altering the transporter's sequence. Thus, one regularly sees mutations that amplify transporter genes in the yeast experiments, and much less frequently mutations that affect the amino acid sequence of the transporters.

Another common event seen in these experimental circumstances is change, by some kind of chromosomal rearrangement, to the *cis*-acting regulatory sequences that direct the transcription of a gene product that is made in inadequate amounts. As a result of such changes, a strong promoter can be substituted for a weak one or for a promoter that does not respond to the repression that is keeping the activity of the normal promoter in check. Such mutations are often found in both the yeast system and in tumors. The example given in Chapter 7 of a recurring reciprocal translocation in cancer (called the Philadelphia chromosome) serves nicely to illustrate this point. This translocation causes most of the coding sequence of a weakly expressed growth factor, which normally limits the growth of a leukemic cell, to come under the control of a promoter that is highly expressed in the leukemic tumor cell type. This gives an already cancerous cell a substantial growth advantage.

A third common event, again found in both the yeast system and in tumor cells, is a deletion. In yeasts, one such deletion removes a transporter that is inhibiting growth under selection. Such deletions are one way in which copies of active, recessive oncogenes (so-called "tumor-suppressor" genes) can be lost, resulting in cells that have lost the regulation that keeps them from becoming tumors.

The reader will, by now, have seen the point. The modular regulatory architecture of genomes, in which the *cis*-acting regulatory sites are separate from the coding (or other functional) sequences, enables the creation of mutant variants that express much more (or much less) of an otherwise unaltered gene product, to produce a large effect on a cell's fitness.

It is important not to generalize too freely from these situations to the overall process of evolution. Both of the examples I have mentioned (yeasts in culture and developing tumors) are special cases that lend themselves to large-step selection. More often, under natural conditions, there is a relatively constant local environment, and the need to respond to continuing changes in the environment is minimal. In the yeast assay, this situation is completely different, as the environment is changing on an hourly basis. Indeed, when the successful, evolved individuals in the yeast assay are tested for relative fitness against their ancestors, they generally are more fit only in the environment in which they were selected for. In other environments, they prove to be less fit; the change that helped them in one environment often turns out to be deleterious in another. Thus, we see the reality: The evolved organism is a compromise that fares well over all, integrating adaptations to each of the environments that the population is likely to experience regularly.

For animals, evolutionary selection occurs on the whole organism rather than on a single cell or cell type. The evolution of a cell into a tumor is therefore not really a proper example of whole-organism selection, nor of the evolutionary changes that would have led to the animal. Although nobody knows for sure, I think it safe to say that when a population's environment changes, new selective pressures can drive the selection of large-scale genomic changes; over the long term, however, an organism's environmental optimization occurs in smaller steps, with gross changes in genomic regulation or protein content less likely to be successful.

## The Domain Architecture of Proteins and Their Genes

Of course, proteins do evolve, and much of this clearly happens via the selection of point mutations, as has become apparent from the huge amount of comparative sequence analysis done in recent years. As we will see in the next chapter, this optimization of protein function is a slow process that is often preceded by the duplication of a gene or genomic locus, which allows even essential proteins to diverge while maintaining the fitness of an organism.

Evolutionary progress is greatly speeded by the fact that proteins, like genomes, turn out to have a modular structure, containing functional "domains" that contribute to the overall biological role of the protein. For example, in Figure 12.2, I have portrayed the domain structure of a protein that is important for the physiology of blood clots in humans. Tissue-type plasminogen activator (t-PA) is a protein that dissolves fibrin clots by acting on plasminogen (a highly similar protein) to activate it by proteolysis; the activated protein is called plasmin. Plasmin, in turn, degrades fibrin (the protein the makes up the clot) into short peptides. t-PA was one of the first important products of the biotechnology industry. Today, fibrinolytic treatment with t-PA, and with other similar proteins,

Tissue plasminogen activator

**Figure 12.2.** Domain structure of human tissue plasminogen activator (t-PA). The brown arrowheads indicate the positions of introns. EGF, epidermal growth factor domain; F, fibronectin domain; K1 and K2, kringle domains; SP, signal peptide; P, protease.

has become a standard part of the treatment for heart attacks and some kinds of strokes.

Figure 12.2 shows the domain structure of t-PA; each domain forms a compact, folded structure partly held together by disulfide linkages between cysteine residues. The domains of this protein, starting at the carboxy (right) end of the protein are, as follows.

- *Protease domain:* At the carboxy (right) end of the protein sequence is a serine protease domain (P), which holds the "active site" that contains the residues that carry out the chemical hydrolysis of peptide bonds. Close homologs of these domains are found in many dozens of other proteases, each with its own substrate specificity that is dictated not only by the structure of this domain itself but also by the other domains to which it is attached. One of these proteases is plasminogen, which degrades fibrin when activated to plasmin. Like t-PA, plasminogen has a protease domain at its carboxy end. The amino acid sequences of the plasmin and t-PA protease domains are strikingly similar.

- *Kringle domains:* The two domains adjacent to the protease domain are called "kringle domains" (K1 and K2). They are named for a Scandinavian pastry that their structure is said to resemble. Kringle domains are found in many of the proteins that act in blood clotting or in clot dissolution. Plasmin has five tandem kringle domains, whereas t-PA has two; all are highly similar in amino acid sequence.

- *Epidermal growth factor domain:* Beyond the kringle domains is an epidermal growth factor (EGF) domain that is homologous to the EGF domain found in numerous extracellular membrane proteins. These growth factor domains are implicated in the binding of cells to extracellular matrices; some proteins have as many as 30 tandem EGF domains.

- *Fibronectin finger domain:* A fibronectin "finger" domain (F) is found in numerous extracellular matrix proteins; fibronectin itself contains a dozen of them.

- *Signal sequence:* At the start of the t-PA polypeptide chain (at the amino terminal, left end) is a signal sequence (SP), which directs the nascent protein to the secretion pathway. t-PA, like all the proteins that contain this domain, must be secreted from the cells that produce it.

The functions of t-PA depend on these various domains. The finger and EGF domains are implicated in localizing t-PA to fibrin clots; the K1 is implicated in stimulating t-PA's protease activity in the presence of fibrin; the signal sequence is essential for secretion from the cell. Thus, the protein is a collection of functional modules with diverse properties that function together to provide protease activity only where and when it is needed to remove an undesired fibrin clot, such as those that contribute to heart attacks and strokes.

The implication of this modular arrangement, from an evolutionary perspective, is similar for both proteins and genomes. Evolutionarily, t-PA is clearly closely related to plasminogen, fibronectin, and EGF, having inherited its domains from the common ancestors of each of these proteins. This illustrates how new protein functions are created in evolution through the reassortment, amplification, and deletion of the DNA sequences that encode the functional modules of existing proteins.

There is another feature indicated in the illustration of t-PA. The small arrows in this illustration show the positions of the introns. Introns are the sequences that separate the coding regions (the exons) of a gene, and they are removed when the primary RNA transcript of a gene is spliced to make the mRNA for delivery to the ribosomes. The presence of an intron at nearly every boundary between its functional domains strongly suggests that the evolution of t-PA, and proteins like it, involved some kind of recombination events in the introns.

As discussed in Chapter 7, some of these events appear to have been a result of homologous recombination occurring between repeated sequences in the genome, whereas others may have been transposon-mediated or completely illegitimate events. When these events occur in introns between domain-encoding exons, splicing mechanisms ensure that the spliced mRNA from the primary transcript of the newly reassorted, novel gene can be translated. The splicing machinery sees the exon boundaries only and puts the exons together, preserving the reading frame.

It should be clear that evolutionary processes at both the genome level and the individual protein-coding gene level involve not only amino acid changes but also the reassortment of the functional bits of sequence, promoting rapid changes in protein function. The presence of dozens of copies of, for example, EGF-binding domains in some proteins and only a few in many others suggests that selection for the ability to bind EGF is stronger in some proteins than in others.

The modular structure of genomes, together with the presence of introns between functional domains, highlights how the amplification and reassortment of these domains can occur relatively easily. This architecture promotes profound and rapid evolution of protein function that dwarfs the speed of evolution solely because of changes in amino acid sequences brought about by point mutations.

CHAPTER 13

# Evolution Conserves Functional Sequences

The genomes of thousands of different species have been sequenced and analyzed in recent years and are available in data repositories. Each of them represents the current state of the evolution of an organism. Comparative analyses of these genome sequences have revealed shared similarities that show, unambiguously, that genomes are the product of the evolutionary process. The most conserved sequences in each of these genomes can be traced back to the common ancestor of all life on this planet (Fig. 13.1). Plant genome sequences are more similar to each other than they are to those of other kinds of organisms, and the human genome is more similar to those of the great apes than it is to the genomes of horses or pigs. Despite their manifest differences, all genomes have features in common that reflect the basic features of molecular biology (i.e., how proteins are encoded in the genome and how their synthesis is controlled) and also of evolution (e.g., the fate of variants that arise in natural populations as they are subjected to natural selection).

It is safe to say that only a fraction of the thousands of genome sequences now available have been fully annotated, but parts of all of them have certainly been used in sequence-comparison studies of different types. The human genome sequence, the mouse genome sequence, and the sequences of the experimentally tractable model organisms and major pathogens have been studied in the greatest detail. It is only for this small subset that thorough annotation has been attempted. To this end, a variety of databases and computational tools are freely available, including sequence browsers that allow anyone to look at these sequences and see general features that apply to all of them. Particularly useful general browsers can be found at http://www.genome.ucsc.edu and at http://www .ensembl.org. The major model organism–specific sites include SGD, the *Saccharomyces* Genome Database (http://www.yeastgenome.org); FlyBase (http://fly base.org/); WormBase (http://www.wormbase.org); and Mouse Genome Informatics (http://www.informatics.jax.org). The tools available at these sites include

**Bacteria**

**Archaea**

**Eukaryotes**

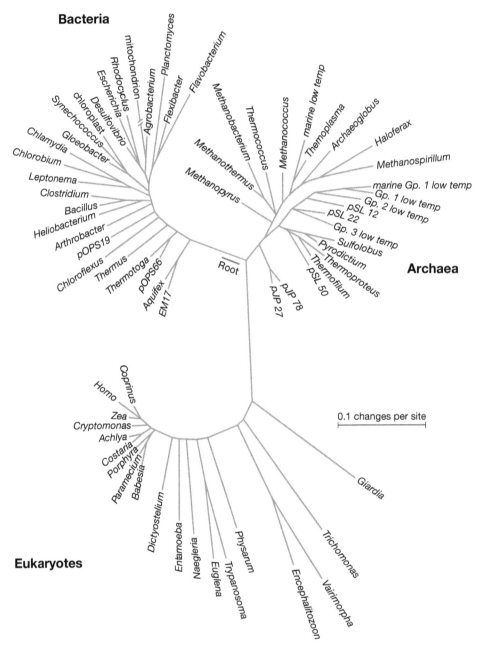

Root

0.1 changes per site

**Figure 13.1.** Evolutionary tree of life. This phylogenetic tree is based on sequence similarity among ribosomal RNA sequences. (Reproduced from Pace NR. 1997. *Science 276*: 734–740, with permission from AAAS.)

a range of analytical tools, for example, to conduct similarity searches, compare gene order between different species, and perform statistics on the prevalence of particular sequences.

The genomes of organisms vary greatly in size. Bacterial genomes are small: The genome of *E. coli* is a little over 4 million bp. Simple methods for detecting ORFs (see Chapter 7) indicate that ~88% of the *E. coli* genome encodes proteins. Yeast genomes are also small: The genome of *S. cerevisiae* is 12.1 million bp, ~74% of which encodes proteins. These organisms grow very rapidly; they can double their numbers in a matter of 0.5–2 h. They grow as single cells, sometimes in colonies or communities. Their genomes are small and largely devoted to encoding proteins.

The genomes of multicellular organisms also vary widely in size. The genome of the nematode worm *C. elegans* is ~100 million bp, but only ~25% of this encodes proteins. The genome of the fruit fly *D. melanogaster* is somewhat larger, ~140 million bp, of which only ~17% encodes proteins. The genomes of mammals (humans or mice) are very much larger (3.3 billion bp), and the fraction of their sequence that encodes proteins is only ~1.1%.

A little back-of-the-envelope calculation shows, surprisingly, that the total amount of sequence dedicated to encoding proteins in all of these genomes is very much more similar than the amount of total sequence would suggest: 25 million bp in worms and flies to as much as 35 million bp for the mammals. If we recognize that the average size of a protein in all of these organisms is very similar (roughly 400 amino acids, which would each require 1200 bp to encode, as each amino acid is specified by a triplet of bases), it would appear that all these eukaryotes must contain a similar number of protein-encoding cistrons (i.e., 21,000 or so for flies and worms and, at most, 30,000 in mammals). In reality, the mammalian number is an overestimate; at the time of writing, it is thought that the human genome contains approximately 22,000 protein-encoding genes.

These numbers suggest very strongly that genome evolution does not simply result in the creation of more and more proteins as organisms become more complex. One logical place to look for non-protein-coding sequences that contribute to fitness is among those that are conserved in evolution. The hypothesis that these sequences might be major contributors to fitness in more complex organisms is supported by some additional genomic statistics. The amount of substantially conserved sequence that is noncoding in bacteria is minimal (no more than a few percent), somewhat higher in yeasts (on the order of 10%), higher still in worms and flies (~40% and ~60%, respectively), and ~80% in humans and other vertebrates. Thus, as organisms become more complex, more and more noncoding sequence appears to be under selective constraint.

## Evolutionary Conservation of Functional Gene Sequences

Most of what we know about the evolution of individual functional sequences we have learned from genes that encode proteins. I will therefore consider first the evolutionary processes that act on coding sequences, and will discuss later why it is that noncoding conserved sequences have come to be so well conserved in more complex organisms.

Similarity among proteins is assessed from their amino acid sequence, rather than from the DNA sequence that encodes them. One reason for this is that the same amino acid sequence can be encoded by many different nucleotide sequences, because of the degeneracy of the genetic code (see Chapter 7). Another reason is that selection is almost always exerted on the functionality of a protein, rather than on the specific codons that encode it.

***Conservation Implies Functionality:*** The modern interpretation of Darwinian evolution underlies all of our molecular analyses of sequences. If we restrict our attention to just those sequences that encode proteins or RNAs, we see a remarkable correlation emerge that extends across the entire tree of life, which represents several billion years of evolution. The proteins that carry out the most fundamental biochemical processes (translation, transcription, and DNA replication) are found to be the most highly conserved in amino acid sequence. This rule was clearly evident in recent experiments carried out in the laboratories of the Australian–American population geneticist Marc Feldman and American geneticist and mathematician Michael Eisen with four closely related species of yeasts.[1] In these experiments, the researchers measured the effect that deletions in protein-encoding sequences had on growth rates and used this as a proxy for their contribution to fitness. They found a negative correlation to exist between the rate of evolution of a sequence (as assessed from comparing it among the four different yeast species) and the effect of its deletion on fitness, even though the most conserved gene sequences (e.g., those that encode the ribosomal RNA and proteins) hardly varied between these close relatives.

***Neutral Theory of Molecular Evolution:*** In 1968, Motoo Kimura introduced the neutral theory, which posited that most of the mutations that are constantly occurring have very little or no effect on fitness. It is these neutral mutations that are responsible for most of the continuing change in sequence over time. The frequency of a neutral mutation in populations increases or decreases by

---

[1] Wall DP, Hirsh AE, Fraser HB, Kumm J, Giaever G, Eisen MB, Feldman MW. 2005. Functional genomic analysis of the rates of protein evolution. *Proc Natl Acad Sci* **102**: 5483–5488.

random drift until it either becomes fixed in the population or is lost. Thus, most of the mutations that are used as a basis for assessing sequence similarity are neither selected for nor selected against.

> ▶ *The simple base-pair changes in sequence that allow us to follow evolution and to construct trees of life are mainly neutral, and do not have a significant effect on the fitness of the organisms that carry them.*

The neutral theory also says that a relatively small fraction of mutations have phenotypic consequences; mostly these are not neutral and instead have significantly deleterious consequences. These mutant alleles are destined to be removed from the population at a rate that depends mainly on the degree of their negative effect on fitness. It is this "negative" or "purifying" selection that guarantees that molecular sequences that encode functions fundamental to survival (such as translation, transcription, and DNA replication) change least rapidly over evolutionary time. Individuals with deleterious mutations in these sequences are subject to the strongest negative selection, and thus they do not persist long in the population.

Finally, of course, a much smaller fraction of mutations improve fitness and thus are subjected to positive selection. For the reasons suggested in Chapter 12, many of these mutations turn out not to be simple changes but rather CNVs (copy number variants) and/or chromosomal rearrangements, reducing even further the number of simple base-pair changes that are positively selected for.

Some uncertainty remains (and not a little controversy) about the possibility that many mutations might not be exactly neutral and that some neutral alleles rise to fixation by mechanisms other than simple random drift. Evolutionary theory continues to be a contentious field, although less so since the increased availability of genomic sequence data from different organisms, which provides a resource for checking theory against observation. Nevertheless, the basic elements of Kimura's neutral theory form the basis of a modern consensus about the major features of molecular evolution. Among its other advantages, the theory also accounts for the naturally high levels of DNA sequence polymorphism that are found in most populations, both in coding and noncoding DNA.

**Duplication and Divergence:** In 1970, Japanese–American geneticist Susumu Ohno proposed a framework for thinking about how those genes that are under strong purifying selection (such as those that encode essential proteins) might nevertheless evolve. He proposed that duplications of such genes might arise, to produce two copies of a single gene (see Chapter 7). One of these copies would continue to serve the function under selection, whereas the other copy would be free to evolve to produce a novel protein that contributes to an organism's overall fitness, even though it might then no longer fulfill its original function. There are many ways in which this new protein could contribute to improved fitness, including changes in protein regulation, localization, catalytic activity, etc. There is ample evidence in today's genome sequences, including in our own, that this process of gene duplication, followed by the functional divergence of one copy, has occurred over and over again.

One classical example of this in the human genome (and in all other mammalian genomes) can be found at the locus where the genes that encode the hemoglobin β-chains reside (one of which is the *HBB* gene, in which the sickle cell mutations discussed in Chapter 2 are found). This locus comprises a tandem array of five very similar sequences that encode very similar proteins, each of which has some properties in common with the others (such as the ability to combine with β-globin chains to make tetramers that bind oxygen and carbon dioxide) and some divergent ones. Among the divergent properties are differences in regulation (each is expressed at different time points in the development of the human fetus) and differences in catalytic properties (which endow hemoglobin tetramers with different affinities for oxygen and carbon dioxide).

A similar tandem array of sugar transporters is found in yeasts, each of which differs from its neighbors in its regulation and catalytic properties. Both the variant globins and the variant sugar transporters clearly arose by duplications. The duplicates ultimately became fixed in the population because the duplication itself and then the divergence into several alternate functional proteins had selective value.

The abundance of genes that clearly are the result of duplication and divergence, sometimes over and over again, means that there are, especially in higher organisms, large families of functional genes that are closely related in sequence. A useful nomenclature was introduced by the evolutionary biologist Walter Fitch in 1971 to describe these family relationships and their inferred evolutionary history.

*Paralog:* "Paralog" is the name given by Walter Fitch to duplicated copies of the same ancestral sequence that have diverged in function but been retained over evolutionary time because each contributes to the overall fitness of the organism as it has evolved.

*Ortholog:* "Ortholog" is the name given to sequences conserved in two different species that each evolved linearly from the same ancestral sequence. If, during this evolution, one or more instances of duplication and divergence occurred, all the paralogs in the evolved species are orthologs of the ancestor, because "ortholog" is defined by sequence similarity and phylogeny and not by function. That said, when there is clearly a single ortholog among all the paralogs found in a sequence similarity search, it usually has an identical or very similar function. Figure 13.2 shows how the mammalian globins evolved by duplication and divergence from a single sequence into the α and β subunits (paralogs of each other) in three vertebrate species. The duplication and divergence of the β-globins into more differentially regulated fetal and adult subtypes makes them paralogs of each other, but orthologs of a common β-globin precursor.

*Homolog:* "Homolog" refers to sequence similarity more generally. Thus, orthologs and paralogs are subsets of homologs.

Duplications come in many forms. In addition to the CNV events discussed in Chapter 7, there is evidence in modern genomes of ancient chromosomal duplications having occurred, and also of a series of whole-genome duplications, as well as other kinds of events that produce more than one copy of sequences in a genome. Any or all of these can serve to provide duplicated genes, the sequence and function of which can then diverge, providing both specialization of function and new functions.

In discussing the modular nature of proteins in Chapter 12, I provided the example of t-PA, which contains a number of different domains shared by plasminogen, fibronectin, EGF, etc. (see Fig. 12.2). Knowing this, it should be clear

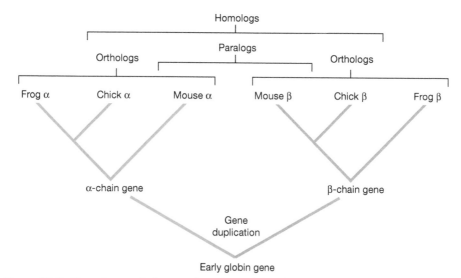

**Figure 13.2.** Homologs, orthologs, and paralogs of the globin genes. This schematic illustrates the evolutionary relationships between the homologs, orthologs, and paralogs of the α- and β-globin genes among three mammalian species.

that homology must be assessed piecewise, domain by domain. Thus, the kringle domains of t-PA are paralogs of the kringles in plasminogen, the finger domain of t-PA is paralogous to the finger domain of fibronectin, and so on.

Selection is applied from the environment whenever organisms reproduce; the types of variation available for selection to act on include copy number changes, translocations, and, of course, point mutations. A common theme concerning the way in which evolution works has emerged from studies of sequence similarity among the earth's organisms, of experimental evolution in model systems, and of cancer genomes. It is this: The architecture of the genome, especially the modular nature of its functional parts, favors amplifications, regulatory variants (where normal regulatory sites are replaced by others), and exchanges of protein domains among duplicated copies of genes. These are the types of large-step changes one sees in experimental systems. Smaller sequence changes that alter the functional domains of proteins to increase fitness also occur, but they appear to be less frequent and their effects more subtle.

## The Phenotypic Effects of Mutations in Conserved DNA

Up to this point, I have focused on the kinds of mutations that appear to have been selected in evolution. I have put forward the view that big changes are usually the result of genomic rearrangements that alter the amount or regulation of proteins. The evolution of proteins and protein domains appears to proceed by the more subtle changes that result from point mutations. In this section, I discuss the effects of mutations in the different regions of genes and genomes that have been conserved over evolutionary time.

*Protein-Coding Sequences:* As I indicated above, these sequences comprise the most obvious, but certainly not the only, focus of natural selection. The amino acid sequences of proteins are clearly subject to purifying selection. This is easiest to appreciate by comparing the frequency of synonymous with nonsynonymous changes of codons. Although it is conceivable, as I indicated in Chapter 7, that synonymous mutations can occasionally result in a phenotype, it is very much more likely that nonsynonymous mutations will do so, and more likely still that nonsense and/or frameshift mutations will produce a phenotype. Indeed, when these latter two kinds of mutations occur early in the coding sequence, they will produce a non-functioning protein, if one at all, and thus a null phenotype.

In experimental evolutionary studies conducted in bacteria or yeasts, one finds, in addition to genomic rearrangements, the occurrence of quite a few point mutations. Many of these base changes, on further investigation, turn out to be neutral, having no detectable effect on fitness in the selective condition to which the cultures have adapted. The ones that can be shown experimentally to increase fitness, and to account for evolutionary change in these conditions, generally turn out to be nonsynonymous codon changes.

A similar pattern emerges in sequence studies of tumor genomes. Here, it is less easy to experimentally test the effect of individual mutations on fitness. Nevertheless, mutations appear repeatedly in the coding sequences of particular genes (called "oncogenes" or "tumor suppressor" genes). Almost without exception, such point mutations are nonsynonymous codon alterations, nonsense mutations, or frameshift mutations.

> ▶ *The repeated, independent finding of nonsynonymous codon changes, nonsense mutants, frameshifts, and/or small deletions in a particular gene's coding sequence is the most frequent kind of evidence that a gene is responsible for a phenotype, including an inherited human disease.*

For the human genome, an easily accessible source of information on what kinds of mutations produce substantial phenotypes is the Human Gene Mutation Database (http://www.hgmd.cf.ac.uk/), which compiles these kinds of data from the literature. At writing, of 100,000 human mutations consisting of small gene lesions (i.e., not CNVs or translocations) that are responsible for inherited disease, ~56,000 (>50%) are missense or nonsense mutations; 16,000 are microdeletions ($\leq$20 bp); 6000 are small insertions ($\leq$20 bp), and 9000 are RNA-splicing defects (see below).

The observation that nearly all (87,000 of 100,000) of the genetic lesions that cause inherited disease in humans affect protein production strongly suggests that more-subtle mutations may only rarely cause a phenotype that follows a simple Mendelian pattern of inheritance in families. This association is also seen in model organism genomes, in which the fraction of conserved sequence that is coding is much higher.

Nevertheless, there are several issues that make geneticists like me reluctant to cling to the conclusion that only proteins really matter. First among these is that there is an obvious bias of ascertainment for disease genes: The ones that segregate as single Mendelian factors that control unambiguous traits are also the first to be found. Second, many gene sequences have been determined from copies of mRNA (or from the complementary DNA made from mRNA) and not from genomic sequence; if no mutation is found in the complementary DNA, genomic sequencing might not have been done or might not have succeeded in finding any mutations—these cases are missing from the databases. Third, it may be that the subtle phenotypes produced by some mutations are more or less dependent on other mutations in the genetic background of an organism, and so never have a simple Mendelian inheritance pattern; instead, they would display more complex modes of inheritance. This last point is a matter of considerable debate as I write this. The possibility that inherited diseases with complex modes of inheritance might be the result of interactions among many genes is a subject I will return to shortly.

***Transcribed Noncoding Sequences in Genes That Encode Proteins:*** The transcription of protein-encoding genes produces a primary transcript that contains sequences upstream and downstream of the coding sequence itself, which are called the 5′ untranslated region (5′ UTR) and 3′ untranslated region (3′ UTR), respectively. In humans, as in most eukaryotes, the coding sequences (exons) are also separated by introns. To create an mRNA from this primary transcript, the introns need to be removed and the exons spliced together. In the human genome, the average number of exons per gene is ~10, and inserted between them are very large introns. Although human proteins, on average, contain ~400 amino acids and thus require only ~1200 bp of coding sequence, the average length of protein-encoding transcripts is ~50,000 bases, and they range in size from a few hundred bases to >2 million bases. So, the average human transcript contains on the order of 40 times more intronic sequence than exonic sequence. Most eukaryote genomes are similar to the human genome, although the budding yeast *S. cerevisiae* has only a few very short (tens of base pairs) introns.

Some, but by no means all, of these sequences are strongly conserved. The coding sequence itself (the sequences of the exons) tends to be the most strongly conserved, with introns being much less well conserved. In evolutionary comparisons among eukaryotic species, the base substitution rate in introns is very similar to the synonymous substitution rate in coding sequence, suggesting that most intronic base-pair changes are neutral and not subject to purifying selection.

Splicing is carried out by another complex of proteins and RNAs called a "spliceosome" (in analogy to the "ribosome"), and it is largely agnostic with respect to the sequence of the primary transcript. However, the spliceosome does have to recognize the exon–intron boundaries, which are encoded by the sequences at the ends of introns. The mechanism of this recognition appears to involve the secondary structure of the precursor (pre-) mRNA, as well as base-pairing to the RNAs that are part of the spliceosome. This mechanism of recognition is reminiscent of the way in which ribosomes bind mRNAs in bacteria.

> ▸ *Mutations that alter or delete intron–exon boundary elements are the mutations that most frequently produce a phenotypic effect, second only to nonsynonymous base substitutions in human disease genes.*

The 5′ and 3′ UTR sequences are, however, not nearly as strongly conserved as coding sequences, but they often contain short sequence motifs that are even more strongly conserved than are coding sequences (when assessed at nucleotide sequence level as opposed to amino acid sequence). This is because the degeneracy of the code makes it possible to have 100% amino acid sequence conservation but much less conservation in nucleotide sequence. The highly conserved nucleotide sequences in the UTR sequences of mRNAs have been implicated in many biochemical processes, including control of translation, transport and sequestration of mRNAs, and mRNA stability and degradation. The activity of a group of small, noncoding RNAs, called micro-RNAs (see below), depends on their ability to bind to some of these conserved elements in the 3′ UTR of their targets.

**cis-Acting Regulatory Sites:** *cis*-Acting regulatory sequences in the genome (such as promoters, operators, enhancers, silencers, and origins of DNA replication) are all essential for the proper regulation of macromolecular synthesis, as I have already discussed in Chapters 11 and 12. As such, one would expect them to come under purifying selection. However, there are other *cis*-acting sequences that I have not discussed (such as the signals that control mRNA activity and decay), which also affect the regulated expression of proteins. Comparative sequence analyses have provided evidence that each of these elements is under purifying selection.

The degree to which these sequences are conserved, however, is quite variable. Many of them appear to be evolving more rapidly than the proteins they control. One explanation for this might be that single base-pair changes in these elements might not always produce severe enough effects to drive conservation to the extremes that one finds for the most well-conserved coding sequences. Another explanation might be that the separateness of these sequences from the coding sequences they regulate allows evolution to proceed stepwise, base pair by base pair, and also by the substitution of one regulatory circuit for another, just like the deletion that resulted in the *lac* operon coming under control of purines (Chapter 11).

**Noncoding RNAs:** There are many RNA molecules, other than mRNAs, that are essential for maintaining the fitness of organisms, and that thus come under purifying selection. First among these are the ribosomal RNAs and the tRNAs. Because these RNAs are absolutely essential to life, it is no surprise that they are among the most conserved sequences of all. However, there are also a variety of other types of noncoding RNAs that are under strong evolutionary constraint.

*Micro-RNAs:* Micro-RNAs (also known as miRNAs) are very short (~22 bases) RNA molecules that posttranscriptionally regulate the expression of their target mRNAs (and possibly of other noncoding RNAs as well). They function by binding to their targets, by base-pairing, to form a short duplex region that is recognized by a complex of proteins that cause the eventual degradation of the target mRNA. They can also inhibit their target's translation. In both cases, the target sequences recognized by miRNAs generally lie in their 3' UTRs, but some functional binding sites have been found elsewhere in mRNAs. These miRNA binding sites might be responsible for much of the conservation of short sequences found in the UTRs.

miRNAs were first discovered in plants, and their significance was elucidated by elegant work in nematodes, in which a key developmental regulatory gene, at first thought to encode a regulatory protein, turned out to encode the precursor transcript of a 22-base miRNA. The posttranscriptional regulation of genes by miRNAs turns out to be a very common phenomenon among eukaryotes, and virtually all produce a set of highly conserved proteins that execute the essential steps in miRNA generation, including the processing of a longer miRNA precursor (which can sometimes contain several miRNA sequences) and the export of these precursors to the cytoplasm. The human genome encodes about a thousand different miRNAs that regulate about half of all human protein-encoding genes.

*Long Noncoding RNAs:* Intensive investigations of transcription patterns in mammals and in other higher eukaryotes have revealed that another class of noncoding transcripts exists: long noncoding RNAs (lncRNAs). These transcripts are not strongly conserved over their length, although most are highly regulated and very often expressed in a tissue-specific manner. Instead, many of them contain small stretches of conserved sequence, particularly sequences that have independently been identified as being enhancers of developmentally regulated genes. There are thousands of these transcripts in mammalian genomes, only a very few of which have been associated with a biological function. One of these is worthy of special mention, because it is responsible for the inactivation of one of the two X chromosomes in female placental mammals, to ensure that females do not have twice as many X-chromosome-derived transcripts as males. The way in which other lncRNAs work is not yet clear, except for speculations that they may be involved in enhancer function. Very few mutant phenotypes have been attributed to loss of lncRNA function.

Although every kind of mutation I have described thus far has been found in most organisms studied, the types of mutations and the kinds of phenotypes (mostly deleterious) they produce are by no means equally frequent. The frequency with which we encounter particular types of mutations and phenotypes depends on the frequency with which the mutations occur (see Chapter 7). In the following chapters, I discuss human genetic mutations, virtually all of which are inferred from relatively strong phenotypes in families or from population-scale studies. In this context, one finds mainly nonsynonymous point mutations, frameshift and nonsense mutations, and splicing defects; followed by CNVs and the occasional miRNA or miRNA-binding mutation; and, even more rarely, lncRNA mutations. As indicated above, cancer genomes are different, because they evolve rapidly and the genetic changes are, in solid tumors, dominated by CNVs and translocations, although point mutations in oncogenes are also important.

## INTRODUCTORY BIOGRAPHIES

**Marc Feldman (b. 1942)** is an Australian-born American population geneticist whose main contributions are in theoretical population biology and the genetics of human populations, especially the transmission of culture along with genes. In more recent years, he has made important contributions to human molecular evolution.

**Michael Eisen (b. 1967)** is an American geneticist and mathematician who transformed the analysis of genome-scale gene expression data by devising a clustering and visualization computer program. Since then, he has been studying the way in which genomes specify form and function and how form and function evolve. Eisen is also a founder of the Public Library of Science (PLoS), an open-access publisher that is transforming the way in which scientists publish their work.

**Motoo Kimura (1924–1994)** was a Japanese theoretical population geneticist responsible for the neutral theory of molecular evolution, which put genetic drift into a foremost position in evolutionary change. Although the neutral theory generated much controversy, it has proved to be a watershed in evolutionary thinking.

**Susumu Ohno (1928–2000)** was a Japanese–American geneticist and evolutionary biologist who theorized that duplication must play a major role in the evolution of genes under purifying selection. His idea of duplication and divergence is now generally accepted. He also suggested that vertebrate genomes are the result of whole-genome duplications.

**Walter M. Fitch (1929–2011)** was an American population geneticist and molecular evolutionist whose contributions include many of the basic mathematical methods of reconstructing phylogenies. His parsimony methods have been incorporated into virtually all modern methods of comparing molecular sequences. He also introduced the important idea of orthology and paralogy, which has made the teaching and discussion of molecular evolution and phylogenies much easier.

CHAPTER 14

# Human Population Genetics

Human biology is not fundamentally different from the biology of other organisms. All species on our planet evolved from a common ancestor, a fact that today we can verify by the straightforward analysis of sequence similarity. I have already emphasized that the basic molecular biology of all living things is virtually the same. Humans are physiologically typical animal species; remarkably, our genomes differ from those of chimpanzees in only one per hundred base pairs. The way traits are inherited in humans is the same as it is not just for animals but for all multicellular eukaryotes, including the pea plants Mendel studied. If there is a basis for thinking our species is biologically exceptional, it lies outside of our basic biology or genetics.

Yet it is impossible to study the genetics of humans in the same way that we study the genetics of flies, worms, yeast, or even mice. The reasons are many, and virtually all of them are obvious. Very basic ethical and moral reasons prevent us from manipulating the mating and reproductive choices of others. Even if we somehow were to get around these social and moral taboos, we would still face two major problems: First, we would find it difficult to achieve statistical significance (e.g., in gene mapping), because we cannot expect to find hundreds or thousands of human progeny of the same cross to study; and second, we would be able to follow traits only through about two generations because the human investigators have the same life span as their human subjects. For these reasons, what we know about human genetics is not based on the kinds of experiments that led to our understanding of genetic principles and mechanisms in bacteria, plants, yeast, worms, and flies.

Instead, the study of human genetics is a process of inference made from observations obtained from existing populations. We are limited to observing the transmission of traits in existing families, examining people in noninvasive ways, or, in some settings, using blood or tissue samples obtained in connection with medical treatment (such as for cancer). Fortunately, much of what we can observe in this passive way is interpretable in light of the ideas and the experimental evidence obtained from experimental organisms. Furthermore, many of our ideas about what might be true of humans can be tested experimentally using suitably designed experiments in tractable model organisms that are not too far removed in evolutionary distance from ourselves, typically the house mouse, *Mus musculus*. This principle provides the foundation for much of modern pharmacology, toxicology, and drug design.

In this chapter, I introduce some basic ideas of population genetics without resorting to the rigorous algebraic exposition that is usually found in textbooks. Instead, I will use simple, back-of-the-envelope estimates. The advantage of this approach is that those readers whose appetite and/or tolerance for equations and formulas is limited should be able to follow the arguments. The disadvantage, of course, is that I will periodically have to warn the reader that these nonrigorous calculations could be misleading if pushed too far. Readers who actually want to make such calculations should consult a textbook of genetics or the original literature, in which the relevant equations and formulas are given in full.

## Gene Frequency and Hardy–Weinberg Equilibrium

In 1908, 8 years after the rediscovery of Mendel's work, Godfrey Harold (G.H.) Hardy, an eminent British mathematician, and Wilhelm Weinberg, a German physician, independently published papers on the mathematical consequences of the rules of Mendelian inheritance at the population level. They pointed out that the frequencies in the population of alleles at a given locus will not change from generation to generation in the absence of mutation, selection, and/or other evolutionary forces, including genetic drift. At first glance, you might think that ignoring evolution is rather drastic, but in reality it is very sensible once one considers the numbers. The rates of mutation are on the order of one base in a hundred million per generation, or less than a mutation per gene per generation. Selection for or against existing polymorphic alleles and genetic drift, despite its crucial role over evolutionary time, is likewise quantitatively minor over the span of just a few generations.

The assumptions, mathematical framework, and language introduced by Hardy and Weinberg became the foundation of population genetics. In the early days of the 20th century, the realization that gene frequencies are basically stable from generation to generation was a major advance in thinking about human genetics.

*The Hardy–Weinberg Law:* The Hardy–Weinberg law, also known as the Hardy–Weinberg equilibrium, provides one of the foundations of population genetics. It assumes that matings in a natural population occur at random with respect to genotypes (other than sex, of course). Consider a polymorphic "autosomal" locus—that is, a locus not located on one of the sex chromosomes, X and Y. This locus, A, has two alleles, A1 and A2. We define the frequencies of these alleles as $p$ (for A1) and $q$ (for A2). Assuming that there are only the two alleles of A (the simplest case), $p + q = 1$. These frequencies can be used to predict the relative frequencies of all the possible diploid genotypes (A1/A1 homozygotes, A1/A2 heterozygotes, and A2/A2 homozygotes) in the next generation. The Hardy–Weinberg assumption of random mating means that the probability of an A1/A1 homozygote will be $p^2$, the probability of an A2/A2 homozygote will be $q^2$, and the probability of an A1/A2 heterozygote will be $2pq$.

There are two straightforward ways of convincing ourselves of this. Because $p$ and $q$ are frequencies, they sum to 1. The binomial expansion also sums to 1, and provides the relative frequency of all the possible diploid genotypes:

$$(p + q)^2 = p^2 + 2pq + q^2.$$

This calculation is easily generalized to deal with multiple alleles and lots of other situations that produce algebraic, but not really conceptual, complications.

**Punnett Square:** A Punnett square refers to an intuitive and equally rigorous, but less generalizable, visualization of the gene frequencies in a population (see Fig. 3.1). It is a diagram introduced by Reginald Punnett, a British geneticist who had gotten G.H. Hardy interested in the issue of gene frequency stability in the first place. The Punnett square has become a standard in introducing Mendelian ratios. If we add up the frequencies in the Punnett square, we once again get $p^2$ A1 homozygotes, $q^2$ A2 homozygotes, and $2pq$ heterozygotes (see Fig. 14.1).

An important thing to note is that the diploid genotypes in the parental generation do not matter. The outcome is the same, for a given $p$ and $q$, whether the diploid genotypes of the parents are distributed at Hardy–Weinberg equilibrium

|  | A1(p) | A2(q) |
|---|---|---|
| A1(p) | A1/A1(p²) | A1/A2(pq) |
| A2(q) | A2/A1(pq) | A2/A2(q²) |

**Figure 14.1.** Calculation of diploid genotype frequencies using a Punnett square. Locus A has two alleles, A1 and A2; their frequencies are $p$ (for A1) and $q$ (for A2), and add up to 1. The Hardy–Weinberg assumption of random mating means that the probability of an A1/A1 homozygote will be $p^2$, the probability of an A2/A2 homozygote will be $q^2$, and the probability of an A1/A2 heterozygote will be $2pq$. See also Figure 3.1.

($p^2 + 2pq + q^2$) or, in the extreme case, are all homozygotes for *A1* or *A2*, a situation that might occur in experimental animals but is highly unlikely in a human population. The essence of the Hardy–Weinberg law is the recognition that in the absence of selection, one always ends up with the same genotype frequencies after a generation of random mating no matter what the distribution of genotypes was in the parental population. This is why some authors refer to this law as the Hardy–Weinberg equilibrium.

Although a full discussion of the role of selection in population dynamics is well beyond the scope of this book, I think the consequences of extreme selection are of general importance. Extreme selection against rare recessive diseases, such as the case in which one of the homozygotes (e.g., *A2/A2*) is effectively lethal, will change the frequencies $p$ and $q$ in every generation, because in each generation some *A2* alleles are lost, reducing $q$. There will be, in such cases, an obvious failure to observe Hardy–Weinberg equilibrium, because in each generation there will be many fewer than expected *A2/A2* homozygotes (and none in the most extreme case). However, even the most extreme selection will not eliminate the *A2* allele from the population quickly, because as $q$ becomes smaller every generation, $q^2$ becomes smaller still, and the fraction of *A2* alleles "hiding," as it were, in heterozygotes ($2pq/q^2$) rises. Over time, the overwhelming majority of the *A2* alleles in the population will be present as heterozygotes. Note also that this is the reason why most of the disease-causing alleles of a recessive disease like cystic fibrosis reside in heterozygotes.

This is why the regrettably popular and dangerous eugenic laws of the last century, aimed at "improving the species" by sterilizing or killing people with rare recessive diseases, are futile. Several hundreds of generations (over literally thousands of years) of extreme selection are required to make only modest reductions in the frequency of a disease allele that is present in 10% or less of the population.

*Inbreeding:* "Inbreeding" refers to the production of offspring by parents who are related to each other by recent ancestry. Inbreeding is sometimes referred to as "consanguinity." I rarely use this latter word, as we no longer believe that blood has anything to do with the passage of genetic information from generation to generation.

Inbreeding is important in human genetics because it greatly increases the probability that a rare, recessive disease allele will become homozygous simply because it was derived from the same recent ancestor.

A simple historical example will suffice to make this point clear. Based on their studies of families, British physician Archibald Garrod and William Bateson (see Chapter 1) inferred, in 1902, that alkaptonuria (the first example of an "inborn error of metabolism") might be inherited as a simple recessive Mendelian phenotype. They noted that many of their families were inbred, which they took as further evidence that Mendel's rules apply to humans as well as to plants and animals, because they understood that inbreeding will increase the probability of the incidence of recessive phenotypes.

The frequency of alkaptonuria in most human populations is about $1/100,000$ ($10^{-5}$). If the causative allele is in Hardy–Weinberg equilibrium, we can say that $q^2 = 10^{-5}$, and thus the frequency $q$ of the causative allele is the square root of $10^{-5}$, which is 0.0032, or 0.32%.

Now consider the extreme case of a sibling marriage in which one of the parents (Fig. 14.2) is a heterozygote. Each of the siblings has a probability of 0.5 of having inherited the causative allele (*A1* in this figure). The probability that both of them inherited it is the product of these probabilities, or 0.25. Their child has a probability of 0.5 of inheriting the causative allele from each parent. Again, the joint probability is 0.25. By multiplying the two joint probabilities together (i.e., that each parent inherited this allele from their common ancestor and the child inherited it from each of his or her parents), we arrive at the overall probability that the child will have alkaptonuria, which is $0.25 \times 0.25 = 0.0625$. This makes it obviously very much more likely that a child of related parents will have an affected child compared with a child of

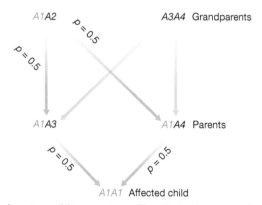

**Figure 14.2.** Inbreeding in a sibling marriage. This illustration shows how the offspring of a sibling marriage can inherit the *A1* allele from each sibling parent and thus twice from the same grandparent. Each sibling has a probability of 0.5 of inheriting *A1*. The probability that both parents inherited *A1* from the same parent is the product of these probabilities, or 0.25. Their offspring has a probability of 0.5 of inheriting the causative allele from each parent. Again, the joint probability is 0.25.

two unrelated parents because the probability of their doing so remains $q^2$, or $1/100,000$.

Inbreeding is significant in practice for the etiology of all rare recessive diseases. If we find alkaptonuria in the child of siblings, it is $6.2 \times 10^{-2}/10^{-5} = 6200$ times more likely that the child has inherited the same allele from his or her grandparent than inherited a different one from the population at random. Note that I have not corrected, in this back-of-the envelope calculation, for the (tiny) probability that an allele other than the one from the common ancestor is nevertheless involved. For first cousins, the same kind of calculation yields $4 \times 10^{-3}/10^{-5}$, which makes it 400 times more likely that their child will inherit this disease allele from them. In this scenario, the possibility that an allele from the general population is involved is closer to making a difference, but it is still negligible. If first cousins have a child with a rare recessive condition, it is overwhelmingly likely that the child is homozygous for an allele that was inherited by both cousins from the same ancestor, in this case a grandparent.

*Autozygosity:* This term refers to homozygosity in an inbred individual that arises as a result of inheriting a stretch of DNA from a recent common ancestor. Autozygosity can be easily detected in diploid genomes by genotyping single-nucleotide polymorphisms (SNPs). This is because in a region of autozygosity one sees only one allele of each SNP; no heterozygous SNPs are expected in the autozygous stretch of genome, which will stretch over millions of base pairs if the common ancestor is a recent one, such as a grandparent.

Although sibling matings are very rare, slightly lower levels of inbreeding are not at all rare in modern human populations. Specifically, there are large regions of the world where cousin-to-cousin and uncle-to-niece marriages (for which the probability of descent from a common ancestor is twice that of cousins) comprise more than half of all marriages.

*Homozygosity Mapping:* Homozygosity mapping involves identifying a region of homozygosity held in common by two or more inbred individuals with a recessive disorder, using SNPs or other polymorphic DNA markers. This approach is based on the expectation that the causative gene will be embedded in a region of autozygosity, and it has been used to locate (map) thousands of rare and recessive disease genes that follow a simple recessive Mendelian pattern of inheritance.

For example, consider two siblings affected by such a disease, who are the result of a marriage between second cousins. Second cousins are the children of first cousins (they share the same great-grandparents); first cousins are the children of siblings (they share the same grandparents). Second cousins have, on average, about $1/64$ of their genomes identical by descent. The probability that they will have the *same* region autozygous is about $1/64 \times 1/64$ (or about $1/4000$). This value is only an estimate, because once again I did not correct for the relatively unlikely possibility that one of the siblings acquired a rare recessive allele from an ancestor other than the common ancestor. That said, $1/4000$ exceeds the minimum threshold for statistical significance that is conventionally used in human gene mapping; the reader may recall from Chapter 3 that a likelihood ratio of 1000 (a LOD score of 3) is the threshold used in that setting.

Of course, it is imprudent to rely only on what amounts to a single observation. It should be clear, nevertheless, that only a handful of inbred affected individuals are required to localize a Mendelian disease gene that produces a recessive phenotype using homozygosity mapping. In contrast, dozens of affected individuals are required when using standard gene-mapping methods to achieve statistical significance.

## Recombination, Haplotypes, and Linkage Disequilibrium

In earlier chapters, I introduced the idea that crossing over between chromosomes occurs during meiosis to account for the observation that loci on the same chromosome recombine. The frequency of recombination ($F_{rec}$; for more on this, see Chapter 3) is a measure of the physical distance between loci; adjacent loci will recombine very infrequently, but loci located at opposite ends of the same chromosome will recombine with $F_{rec} = 0.5$. Today, with modern DNA technology, we can assess the presence of DNA polymorphisms by the millions, allowing us to follow the process of recombination in detail, to high resolution.

Let us follow meiosis and recombination in an individual who inherited chromosomes from his parents, which differ from each other at hundreds of thousands of SNPs. I have denoted the maternal alleles of each SNP with an M and the paternal alleles with a P, and the centromere with an asterisk. I have used the asterisk to show attachment of the two chromatids as well. The homologous chromosome pair synapsed in the first stages of meiosis will look like this:

MMMMMMMMMMMMMMMMMMMMMMMMMMMMMMMMM * MMMMMMMMMMMMMMM    (1)
MMMMMMMMMMMMMMMMMMMMMMMMMMMMMMMMM * MMMMMMMMMMMMMMM    (2)
                                 *
PPPPPPPPPPPPPPPPPPPPPPPPPPPPPPP * PPPPPPPPPPPPPP    (3)
PPPPPPPPPPPPPPPPPPPPPPPPPPPPPPP * PPPPPPPPPPPPPP    (4)

In most eukaryotes, meiosis results in about one or two crossovers per chromosome arm. For this example, let us imagine two crossovers happen: one between strands (1) and (3) on the left arm and the other between strands (2) and (3) on the right arm. The recombined chromosome pair now looks like this:

PPPPPPPPPPPPP MMMMMMMMMMMMMMMMM * MMMMMMMMMMMMMMM    (1)

MMMMMMMMMMMMMMMMMMMMMMMMMMMMMMMM * MMMMMMMM PPPPPP    (2)

*

MMMMMMMMMMMMMM PPPPPPPPPPPPPPP * PPPPPPPP MMMMMMM    (3)

PPPPPPPPPPPPPPPPPPPPPPPPPPPPPP * PPPPPPPPPPPPP    (4)

The gametes (sperm or egg) from this meiosis will contain one of these four recombined chromosomes:

PPPPPPPPPPPPP MMMMMMMMMMMMMMMMM * MMMMMMMMMMMMMMM

(derived from strand (1) above), or

MMMMMMMMMMMMMMMMMMMMMMMMMMMMMMMM * MMMMMMMM PPPPPP

(derived from strand (2)), or

MMMMMMMMMMMMMM PPPPPPPPPPPPPPP * PPPPPPPPMMMMM

(derived from strand (3)), or

PPPPPPPPPPPPPPPPPPPPPPPPPPPPPP * PPPPPPPPPPPPP

(derived from strand (4)).

Between every generation and the next, there is a round of meiosis and crossing over. The result is that, in effect, the DNA in chromosomes is passed on in the form of continuous DNA segments, the average length of which is half of a chromosome arm, or about 100 million bp. With each generation, the stretch of chromosome that carries a particular ancestral sequence of SNP alleles becomes shorter, as we can visualize if we consider the next meiosis, in which the chromosome of one of the recombinant gametes is paired with another that looks like one of the original parents (i.e., gametes (1) and (4) from the above meiosis):

PPPPPPPPPPPPP MMMMMMMMMMMMMMMMM * MMMMMMMMMMMMMMM    (1)

PPPPPPPPPPPPP MMMMMMMMMMMMMMMMM * MMMMMMMMMMMMMMM    (2)

*

PPPPPPPPPPPPPPPPPPPPPPPPPPPPPP * PPPPPPPPPPPPP    (3)

PPPPPPPPPPPPPPPPPPPPPPPPPPPPPP * PPPPPPPPPPPPP    (4)

When this individual reproduces, let us imagine that strand (1) recombines with strand (3), and strand (2) recombines with strand (4). The recombined chomosomes in this case would be

PPPPPPPPPPPPPPPPPPPPPP MMMMM * MMMMMMMMMMMMMMM (1)

PPPPPPPPPPPPP MMMMMMMMMMMMMMMM * MMMMMM PPPPPPPP (2)

*

PPPPPPPPPPPPP MMMMMMMMMM PPPPP * PPPPPPPPPPPPPP (3)

PPPPPPPPPPPPPPPPPPPPPPPPPPPP * PPPPP MMMMMMMMM (4)

The four gametes that could be produced would then be

PPPPPPPPPPPPPPPPPPPPPPPP MMMMM * MMMMMMMMMMMMMMM

(derived from strand (1) above), or

PPPPPPPPPPPPP MMMMMMMMMMMMMMMM * MMMMMM PPPPPPPP

(derived from strand (2)), or

PPPPPPPPPPPPP MMMMMMMMMMM PPPPP * PPPPPPPPPPPPPP

(derived from strand (3)), or

PPPPPPPPPPPPPPPPPPPPPPPPPPPPP * PPPPPP MMMMMMMMM

(derived from strand (4)).

Note that the stretches of M alleles and P alleles have been shortened, so that in the next generation the stretches of SNP alleles that are identical to those in the original ancestors have become shorter.

After 25 human generations, the average stretch of contiguous M or P alleles is reduced to about 2 million bp from about 100 million bp. This is still more than enough to be able to detect a stretch of contiguous SNP alleles that have stayed together, as current technology allows the simultaneous assessment of a million SNPs, which on average is one SNP locus every 2000–5000 bp.

In our previous discussion of human gene mapping (Chapter 3), I introduced the word "haplotype," which refers to a set of polymorphic alleles that are inherited together. Clearly the stretches of M or P alleles that have not yet been separated from each other are haplotypes, and they continue to be referred to as such.

After a much larger number (effectively an infinite number) of generations, each SNP will have been separated from its neighbors by recombination. No vestige of the original maternal linkage will remain, the M and P alleles will be arranged randomly, and there will no longer be any detectable surviving ancestral haplotypes.

One of the most important realizations to come out of the study of human populations using DNA technology is that this limit has not yet been reached for most human populations. Much of our genome consists of haplotype blocks, some of which cover millions of base pairs (e.g., the regions around centromeres, where recombination is inhibited), and a few of which appear to have been inherited as far back as the earliest humans. The median size of haplotype blocks in the human genome is ~20–50 kbp, corresponding to 0.00006% of the genome.

Thus, if we take into account linkage disequilibrium and mark each haplotype block with DNA polymorphisms, we can follow haplotype inheritance patterns pretty comprehensively with a few million SNPs. Today, this can be done using microarrays that are commercially available, making the analysis of tens of thousands of individuals affordable.

*Identity by Descent:* Identity by descent (or IBD) refers to genomic segments that are identical because they derive from the same ancestor, even when that ancestor might be very many generations removed. In modern times, IBD is discernible from the patterns of SNP-based haplotypes that have been inherited together over many generations.

However, unlike the detection of autozygosity, where one looks just for stretches in which all the SNPs are homozygous, in population-level assessments of IBD, one has the problem that most individuals are not homozygous. Most convenient existing technologies for assessing DNA polymorphisms (including short-read sequencing methods) only tell you whether an individual is homozygous or not for each SNP. Specifically, these methods do not provide information about linkage phase (see Chapter 3).

Thus, the following situation, derived from my example above, could easily arise. After a few generations, a descendant of the original parents could have a diploid genotype in which two reasonable possibilities are not distinguished:

MPMPMPMMPP    and    MMMMMMMMMM

MMPMMMPMMM            MPPPMPPMPP

It should be clear from the preceding discussion that it is the second possibility that would be interesting with respect to IBD, because it retains the ancestral sequence of 10 contiguous maternal SNP alleles intact. In the example above, there are four haplotypes, only one of which (MMMMMMMMMM) preserves the ancestral sequence. This would occur where there have been insufficient recombination events over the number of generations from the ancestor to the descendant to separate the 10 SNP loci from one another. This haplotype will have been transmitted intact and can be followed in subsequent generations, until recombination intervenes.

*Linkage Disequilibrium:* "Linkage disequilibrium" is the term used to describe the correlation between polymorphic alleles in an ancient haplotype that remains in a population when there has been too little recombination to separate them from each other. As will become clear in the next chapter, linkage disequilibrium plays an important role in detecting and mapping genetic determinants of complex human traits and diseases. Until DNA sequencing technology improves such that we can sequence single molecules over millions of base pairs, linkage phases and haplotypes in humans have to be inferred by statistical methods. These measures of linkage disequilibrium, and the haplotypes that can be inferred from them, have become the basis for mapping the loci that contribute to some of the most common human diseases, such as heart disease, rheumatoid arthritis, and diabetes, as I discuss next.

*Genome-Wide Association Studies:* The genome-wide association study (GWAS) is a population-based method for finding genotype–phenotype associations. Today, this is the only method that has achieved widespread success in identifying genes that might contribute to complex traits. The GWAS method takes advantage of the ability to follow the inheritance of much of the genome by genotyping a few million SNPs.

The basic idea is simple: A large number (usually thousands) of "cases" (e.g., people with a phenotype of interest) are genotyped for a million or more SNPs. The results are compared with a comparably large set of "control" individuals, ideally ones matched for age and ethnicity. Then one looks for an enrichment of SNP alleles among the cases, using statistical methods that correct for the very large number of loci being tested. Any SNP allele that rises to statistical significance (i.e., is found more frequently in the cases than might be expected by chance) is a candidate for marking a haplotype block in which there might be a causative mutation that predisposes to the phenotype of interest. For a simple study, the threshold for significance is a $P$ value on the order of $5 \times 10^{-8}$.

The ideas that underlie the methods used to analyze the result of GWASs derive from theories developed to understand the inheritance of quantitative traits. The underlying assumption is that many genes contribute to a quantitative trait in a way that can be approximated by assigning each locus a fraction of a trait's phenotypic variance.

*Quantitative Trait Locus:* The term "quantitative trait locus" (QTL) is applied to the different loci that together contribute to and determine a quantitative

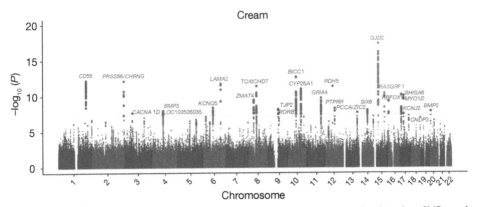

**Figure 14.3.** Manhattan plot for a GWAS for predisposition to myopia. In this plot, SNPs and chromosomes are arranged in map order, from chromosome 1 to 22, on the x-axis. Significance values (the negative logarithm of the P value) are plotted for each of the millions of SNPs on the y-axis. The threshold value is $P = 10^{-7}$. SNPs that rise above this threshold might be the causative mutation or linked to it. (Reproduced with permission from Verhoeven VJ, et al. 2013. *Nat Genet 45:* 314–318, with permission from Macmillan Publishers Ltd.)

phenotype (or trait; see also Chapter 10), such as blood pressure, height, or body mass index. The analysis attributes to each QTL two parameters of interest. One of these is a *P* value, which represents the probability that the locus was enriched in cases by chance. The threshold for significance, in an otherwise uncomplicated genome-wide analysis, is set to about $5 \times 10^{-8}$, at which level the probability of a false positive should be about 5%. The other parameter is the magnitude of each QTL's contribution to the phenotype, often called the "effect size."

***Manhattan Plots:*** Displays of the results of GWASs are called "Manhattan plots" because of their visual similarity to the Manhattan skyline. Figure 14.3 shows such plot for a GWAS for predisposition to nearsightedness (myopia). In this plot, the chromosomes are laid out horizontally in map order from chromosome 1 to 22, and points are plotted for each of the millions of SNPs. Significance values are plotted vertically.

The SNPs that rise above the threshold (presented as a horizontal line at the threshold value of $P = 10^{-7}$), might do so for two reasons. First, they might actually be the causative mutation, if (and only if) this mutation is frequent enough in the population to have been chosen as a SNP for the microarray. Much more likely, however, is that they are linked to a causative mutation that is much rarer in the population *and* that occurred in the past in the background of an already established haplotype block. In this very likely scenario, the SNPs that represent the haplotype blocks closest to the causative mutation will be enriched just as surely as if one of the SNPs is itself causative. So a GWAS is fundamentally yet another way of using linked DNA polymorphisms to map a disease locus.

This means that once a locus or small chromosomal region has been implicated by a GWAS, it remains necessary to find and confirm it as a causative mutation in family linkage studies. One has to sequence the genomic region in affected individuals, find potential causative mutations, and evaluate them by all the same means (often including introducing the causative mutation into the mouse genome; see below) before any confidence can be had in this association.

At the time of writing, there is no robust theory that accounts for how so many genes of small effect found in the population can produce an associated phenotype in affected individuals. There is plentiful evidence that when GWAS is applied to phenotypes with relatively simple genetics (e.g., a Mendelian gene that had yet to be found), it works fine, and locates the right gene or genes. It also finds genes of major effect in complex diseases, such as the major histocompatibility locus (*HLA*) that controls immune response, which comes up very strongly in studies of rheumatoid arthritis, as do other genes that encode proteins previously associated with the disease phenotype. In Figure 14.4, I have provided a Manhattan plot from a study of rheumatoid arthritis in the Japanese population.

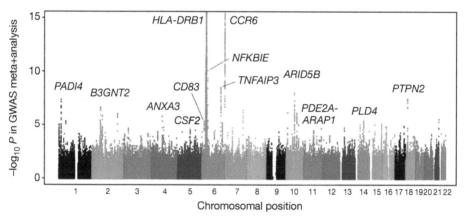

**Figure 14.4.** Manhattan plot for a GWAS for rheumatoid arthritis. This GWAS was conducted in a Japanese population. SNPs are shown in chromosome order (*x*-axis), and the negative logarithm of the *P* value is shown on the *y*-axis. The threshold value is $P = 10^{-7}$. SNPs that rise above this threshold might be the causative mutation or linked to it. A particularly strong peak is seen for SNPs associated with the major histocompatibility gene *HLA-DRB1* and with the chemokine receptor gene *CCR6*. Both are involved in the immune response, which is already implicated in rheumatoid arthritis. (Reproduced with permission from Okada Y, et al. 2012. *Nat Genet 44:* 511–516, with permission from Macmillan Publishers Ltd.)

The general case today, however, is that a GWAS finds very many genes of relatively minor effect. For example, for the phenotype of high blood lipids, 95 loci rose to significance in a recent combined analysis, many of them with plausible links to lipid metabolism.

For now, the interpretation of the significance of one or another SNP associated with a complex, inherited disease is still murky. Whereas a defect in genes of major effect, like the *LDLR* gene, which encodes the low-density lipoprotein receptor (as discussed further in Chapter 15), has real predictive value and mandates any available intervention (in this case cholesterol-lowering drugs), most of the loci uncovered in GWASs have only small effects on the phenotype, even though many of them may have plausible roles in disease biology.

## INTRODUCTORY BIOGRAPHIES

**Godfrey Harold (G.H.) Hardy (1877–1947)** was a British mathematician who made a major contribution to number theory and analysis. He considered himself a "pure" mathematician, although his work turned out to have wide applications in genetics (such as the Hardy–Weinberg equilibrium), as well as in quantum mechanics and thermodynamics. He was a great intellectual leader who is widely credited with bringing rigor to British mathematics.

**Wilhelm Weinberg (1862–1937)** was a German physician with a mathematical flair and an interest in Mendelism who derived the Hardy–Weinberg law independently of Hardy in the same year (1908), although this was not recognized by the English-speaking world until several decades later. By studying families with albino children (a recessive trait), he observed an excess of albino offspring that he correctly attributed to ascertainment bias.

**Reginald Crandall Punnett (1875–1967)** was a British geneticist who wrote what is probably the first textbook on the subject. He cofounded the *Journal of Genetics* with Bateson in 1910. It was he who persuaded G.H. Hardy to take up the issues that led to the Hardy–Weinberg law. The Punnett square is still used to teach Mendelian ratios.

**Archibald Garrod (1857–1936)** was a British physician who introduced the idea of "inborn errors of metabolism." He was the first to realize that alkaptonuria, cystinuria, and albinism were caused by simple recessive Mendelian factors. He was also the first to realize that diseases were more frequent in children of consanguineous parents, and that this made a convincing argument for Mendelian inheritance.

CHAPTER 15

# Inferring Human Gene Function from Disease Alleles

In the very first chapter of this book, I introduced the concept that "genetic analysis relates genotypes to their phenotypic consequences and vice versa." Nowhere is this idea more important than in the interpretation of human DNA sequences. The human genes and loci for which we can most confidently identify a biological function are those for which we can rigorously associate variation in genotype with variation in phenotype. As humans, we are of course most interested in those phenotypes that strongly affect our lives. Thus, much of the interpretation of human genome sequence is based on the study of genes that contribute to disease phenotypes.

Our knowledge of the association of human genes with phenotypes (usually disease phenotypes) is summarized in the continually updated Online Mendelian Inheritance in Man (OMIM) database (http://www.ncbi.nlm.nih.gov/omim). At the time of writing (December 2014), OMIM reported 4308 genes in which a phenotype-causing mutation is known. If the total number of human genes is ~22,000, then we understand the genotype–phenotype relationship for just <20% (4308 of 22,000) of the genes that cause human phenotypes.

Our understanding of genetic function based directly on human variation is, of course, informed by prior knowledge of the biological functions of genes and proteins. This information derives, via the evolutionary conservation of genomic sequences, from other organisms, mainly the genetic model organisms: bacteria, yeast, flies, worms, zebrafish, and, especially, mice. If one can find the ortholog of a human gene (which is not always so simple because of the abundance of paralogs) in a model organism, one can often infer much about the gene's probable function in humans from what is known of its ortholog's function. It is important to note that this information almost always depends on genotype–phenotype correlations in the model organism, which in turn usually depends on the genetic analysis of mutants in that system.

Finally, inferences about whether a candidate gene for a disease is in fact responsible for that disease or for another phenotype can often be tested in model systems. This is most usually done in the mouse by constructing mutations in the orthologous mouse gene and observing the phenotype it produces. In most cases, two kinds of mutations are made: a simple deletion (called a "gene knockout"), which produces the null phenotype, and a sequence alteration identical to the one found in the disease alleles of the human gene (called, by analogy, a "knock-in").

## Simple Mendelian Disease Phenotypes

Many diseases are inherited as simple Mendelian traits. As described in Chapters 3 and 14, most of these genes were found by linkage mapping using DNA polymorphisms, which serves to locate the loci responsible for a Mendelian trait to a resolution of about 1 cM (or 1 million bp). The subsequent localization of the exact locus requires more effort, which I will not describe here except to say that linkage disequilibrium (see Chapter 14) in most populations allows disease-associated haplotypes to be identified in apparently unrelated affected individuals that derive from a common ancestor. Their identification often facilitates the location of the genomic position of the causative Mendelian locus.

Once a disease-associated genomic region has been defined by genetic analysis, sequencing it usually results in the discovery of mutations. Because of the level of variation in our genomic DNA sequence (mostly neutral changes that have minimal phenotypic consequences), considerable effort is required to distinguish disease-causing mutation(s) from adventitious "passenger mutations" in a haplotype. Causative mutations are sometime obvious (e.g., nonsense or frameshift mutations), but sometimes not at all obvious (e.g., missense mutations are often neutral, and regulatory changes are still very hard to find and prove). In either case, the inference has to be checked by making a similar mutation in a model system to see whether it produces a plausibly related phenotype there.

*Recessive Disease Phenotypes:* A well-studied example is the DNA-binding domain encoded by the *TP53* gene. The normal function of the protein encoded by the *TP53* gene is crucial in deciding cell fate when the cell has suffered DNA damage, especially the decision between cell cycle arrest (during which time DNA repair systems are induced by TP53 binding to the promoter regions of repair genes) and cell death (apoptosis, particularly when the damage is irreparable). The loss of this tumor-suppressor gene occurs in many kinds of tumors. Most lung, colon, bladder, and ovarian cancers have mutations in *TP53*, because its normal function is inimical to the growth of evolving tumors.

The TP53 protein is a multidomain protein; one of the domains is a DNA-binding domain, which is essential to its normal function of deciding cell fate after DNA damage. Most *TP53* mutations found in tumors are in the stretch of the gene that encodes this domain. They resemble closely the types of loss-of-function mutations we are familiar with from work on recessive mutations in bacteria, yeast, flies, and worms. Such mutations include nonsynonymous amino acid substitutions, nonsense and frameshift mutations, as well as small deletions.

Figure 15.1 shows the distribution of mutations found in the *TP53* gene. Note the hot spots of missense mutations, which are highly reminiscent of Seymour Benzer's results with the *rII* locus of phage T4 (see Chapter 6). I cannot help pointing out that this compendium of *TP53* mutations is the work of literally thousands of scientists and the expenditure of on the order of tens, if not hundreds, of millions of dollars, whereas the T4 *rII* work was done by a single individual with

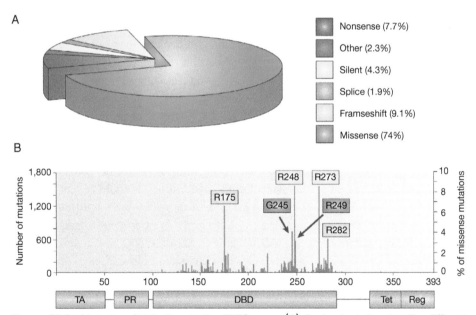

**Figure 15.1.** Mutation distribution in the *TP53* gene. (A) A pie chart showing the different mutation types found in the human *TP53* tumor-suppressor gene reported in the IARC TP53 Database. (B) The distribution of reported missense mutations along the 393-amino-acid sequence of p53. The six most common hot-spot mutations are highlighted in yellow for DNA-contact mutations, green for locally distorted mutants, and blue for globally denatured mutants. The domain architecture of p53 protein is aligned below. TA, transactivation domain; PR, proline-rich domain; DBD, DNA-binding domain; Tet, tetramerization domain; Reg, carboxy-terminal regulatory domain. (Reproduced from Brosh R, Rotter V. 2009. *Nat Rev Cancer 9:* 701–713, with permission from Macmillan Publishers Ltd.)

some technical help at a likely cost of one million dollars or even less. There is no doubt that the effort is worth it to understand our own genes, but the principles, which are conserved over evolutionary time, are much easier to discover and document in model systems.

> ▸ *For simple recessive phenotypes, one should typically find loss-of-function mutations.*

There are important human gene phenotypes that illustrate that recessiveness, even of null alleles, is not always entirely simple. I introduced one such case, sickle cell anemia, earlier in this book (see Chapter 2). Another notable case, and an important issue in human health, is familial hypercholesterolemia (FH). This disease is caused by mutations in the gene that encodes the low-density lipoprotein receptor (LDLR), a cellular membrane protein that is essential for cholesterol transport. Individuals who are homozygous for *LDLR* gene mutations are very seriously affected from early childhood; they have very high levels (6 to 10 times that of normal) of circulating LDL cholesterol. They have severe coronary artery disease and other manifestations, and often have heart attacks in childhood. Many of the more than 1000 mutations that have been sequenced in the human *LDLR* gene are null mutations, including missense mutations, nonsense mutations, deletions, insertions, and splicing defects, which are found all over the coding regions and beyond. Not surprisingly, many of these homozygotes contain no detectable LDLR protein.

Nevertheless, heterozygotes (including the parents of homozygotes) are also affected, even though their cells produce ~50% of the normal amount of LDL protein. Their circulating LDL cholesterol is elevated about twofold above normal, not nearly as much as that of homozygotes, and they also have coronary artery disease, and heart attacks beginning in adulthood. The population frequency of null or null-like mutations in the *LDLR* gene is about 1/500, and heterozygotes account for about 5% of all heart attacks in individuals below the age of 60. Fortunately, the treatment of heterozygotes with statins (drugs that inhibit the biosynthesis of cholesterol) effectively normalizes their LDL cholesterol levels.

American geneticist and cell biologist Michael Brown and American human geneticist and biochemist Joseph Goldstein discovered and thoroughly documented the connection between LDLR and the FH phenotypes. In a recent retrospective, they mention that they pursued FH in part because the incomplete recessiveness phenotype suggested to them that the cause of FH might not be an enzyme, which would be unlikely to be so sensitive to a 50% reduction in level. Transport of cholesterol is not a simple catalytic process, and half the concentration of receptor results in half the rate of cholesterol uptake.

This is another example that highlights that dominance and recessiveness must always be assessed relative to a phenotype. Just as the dominance of $HBB^S$ alleles is assessed differently depending on whether one is looking at anemia, sickling, or resistance to malaria (see discussion in Chapter 2), so the dominance of a null *LDLR* mutation is assessed differently depending on whether one is looking at cholesterol transport, LDL blood level, or predisposition to heart attacks.

**Dominant Disease Phenotypes:** The advent of linkage mapping in humans resulted in the identification of many disease genes that are inherited as dominant traits. In this case, it is not so easy to predict, even from model systems, what kinds of mutations one might expect to find. There are also too many different kinds of mutations to allow their individual enumeration in a short book like this. However, three cases deserve special attention because they exemplify disease-causing mechanisms that appear to be quite prevalent in genes that contribute to common human disease.

**Trinucleotide Repeat Alleles:** One of the classic dominant human disease phenotypes is, of course, Huntington's disease. The mapping, by linkage, of this disease locus was one of the very first successes of human linkage mapping using DNA polymorphisms. Nancy Wexler, an American geneticist, collected samples from large Venezuelan families in which this disease segregated. These samples enabled the human geneticist James Gusella to detect linkage using a DNA polymorphism (an RFLP; see Chapter 1). After this initial success, it took another 10 years to actually isolate the DNA of the causative gene, now called *HTT*. When this gene was sequenced, it was revealed that the dominant mutant alleles of *HTT* had an expansion of a trinucleotide repeat (CAGCAGCAG...). In unaffected individuals, this repeat encodes a run of ~25 (or fewer) glutamine repeats. In affected individuals, it encodes 40 CAG repeats or more. It appears that long runs of polyglutamine cause the huntingtin protein to aggregate, which constitutes a gain of function.

There are other examples in which an expansion of a repeated sequence creates dominant alleles that cause neurological disease. Like Huntington's disease, most forms of spinocerebellar ataxia are caused by glutamine repeats in any of about six different genes residing on different chromosomes. Remarkably, virtually all of these diseases feature primarily neurological phenotypes, and virtually all are associated with some kind of protein aggregation, presumably driven by polyglutamine runs. There are also nucleotide expansion mutations in other genes that appear not to involve the translation of the repeated sequence, such as in fragile X syndrome (which causes mental retardation), Friedreich's ataxia, and myotonic dystrophy. The disease-causing mechanism in these cases is not understood. But whatever it might turn out to be, it will somehow have to involve a gain of function.

Toward the end of the 20th century, it became clear that cancer is a genetic disease, in which somatic cells suffer a series of mutations that allow them to evade biological regulation that normally limits growth. I have discussed the evolutionary implications of this in Chapter 12, emphasizing the influence of genomic architecture in the evolution of tumors. Nevertheless, it is clear that the initial events are, almost always, mutations, and over the years some specialized words have been coined for these.

*Oncogene:* This word refers to a gene that can cause cancer. The first oncogenes were recognized because they were carried in animal viruses that caused cancer. Eventually, cancer biologists succeeded in isolating DNA sequences from tumors. When introduced into cells in culture, these sequences "transformed" normal cells into tumor cells that caused tumor formation when transplanted into animals. Genetic analyses of the tumor virus–derived and of the cell-derived oncogenes revealed that these sequences all came from normal, cellular genes (called "proto-oncogenes"). Typically, these proto-oncogenes encode proteins that normally function in cell growth and differentiation. Often (but not always), the oncogenes in the viruses were mutant forms of the cellular proto-oncogenes; the proto-oncogenes, when introduced into normal cells, usually would not cause tumors, whereas the oncogenes would. Thus, the usage has become quite general that oncogenes are mutated normal genes that can act in a dominant fashion to transform normal cells into cells capable of making a tumor in an animal.

*Tumor-Suppressor Gene (or Recessive Oncogene):* This phrase refers to a gene that can, by loss of its function, cause cancer by transforming a normal cell. In 1971, American geneticist and cancer biologist Alfred Knudson hypothesized how null mutations could cause a dominant pattern of inheritance. He proposed that, during their lifetime, heterozygous individuals lose their unmutated, wild-type allele, either by loss of a chromosome or by mitotic recombination, gene conversion, or even acquisition of a new mutation in the unmutated allele. Any of these events would result in a cell losing its heterozygosity in favor of the loss-of-function allele. If this loss of function results in some kind of a somatic growth advantage, then a cancer (or a cyst) can develop.

*Mutations Affecting Signal Transduction and Allosteric Interactions:* A remarkably large fraction of all solid tumors have acquired an oncogenic mutation in a proto-oncogene called *KRAS*, one of three specific missense mutations. *KRAS* encodes a small member of a very large family of proteins called the GTPases. It normally performs essential functions in signal transduction and/or growth control. KRAS transduces signals by undergoing conformational change when a growth signal is present; this change is accompanied by the hydrolysis of GTP and is normally induced only when the growth of cells is appropriate. Virtually all the *KRAS* mutations found in cancers are specific changes in one of only three codons (usually codon 12, rarely codon 13, and very rarely codon 61). Codons 12 and 13 encode glycine, which in tumors is found to be mutated to glutamate (in the case of codon 12) or arginine (in the case of codon 13). The altered amino acid at position 12 or 13 causes the mutant protein to adopt stably the conformation that normally exists only when KRAS interacts with its partners in the signaling pathway. Thus, the mutant protein signals constitutively, promoting the growth of tumor cells all the time and not just when a growth signal is present. Clearly, this is a gain of function, and requires both a very specific mutant change and strong selection in the somatic cells in which it occurs, because the specific mutations involved should occur very rarely. The cancer environment provides such very strong selection. Many dominant mutations in signaling pathways work by a similar mechanism.

***Dominance via Loss of Heterozygosity:*** This mechanism for dominance in inherited predispositions to cancer, as discussed earlier, was first suggested by Alfred Knudson to explain the highly penetrant, dominantly inherited predisposition to retinoblastoma in newborn children. It turns out to be a very general mechanism whereby alleles of tumor-suppressor genes that cause a loss of function at the cellular level are nevertheless dominant when followed at the level of the whole organism from generation to generation.

Polycystic kidney disease is a relatively common inherited disease that shows simple Mendelian inheritance as a dominant trait in humans. It causes affected individuals to develop numerous cysts in their kidneys, which can become life-threatening with time, although many affected individuals survive to old age. Early linkage-mapping studies identified it as having genetic heterogeneity, resulting ultimately in the identification of two unlinked genes, *PKD1* (on chromosome 16) and *PKD2* (on chromosome 4). Mutations in either of these genes cause essentially the same disease phenotype. The severity of the phenotype is very variable, but the penetrance is high: Affected individuals as confirmed by their genotype almost always develop cysts in their kidneys.

Hundreds of independent germline mutations in human *PKD1* and *PKD2* have been discovered. A remarkably large fraction of these are null mutations, judging from DNA sequence data. For example, at the time of writing, the PKD database maintained at the Mayo Clinic in the United States (http://pkdb.mayo .edu/) lists 203 different nonsense mutations and 310 frameshift mutations in *PKD1*, located all over its coding sequence, and 40 different nonsense mutations and 61 frameshift mutations located all over the coding sequence of *PKD2*. There are deletions and insertions in both genes—every kind of change disastrous to the function of a gene. Homozygous deletions of the mouse orthologs of both *PKD1* and *PKD2* cause embryonic lethality. Thus, individuals affected with polycystic kidney disease are necessarily heterozygotes. Nevertheless, when the cystic tissue from *PKD1* or *PKD2* heterozygotes (see above) was sequenced, it was found that they had indeed lost their heterozygosity for these genes.

Knudson first proposed this idea in the context of retinoblastoma. When RFLPs were first introduced, one of their first applications was to show that loss of heterozygosity is a very prevalent mechanism in the genesis of this cancer. Today, there are literally hundreds of tumor-suppressor genes that are known to undergo loss of heterozygosity to expose an otherwise recessive (at the cellular level) mutation that results in loss of growth control. The gene responsible for retinoblastoma, like most of the others, is a regulatory gene that controls growth—it is the loss of this regulation that results in the initiation of a tumor.

*Dominance via Genomic Rearrangements:* Genomic rearrangements provide another mechanism for the generation of dominant mutations. In the discussion of the regulatory architecture and its implications for evolution (in Chapters 12 and 13), I emphasized that genomic rearrangements can significantly alter gene regulation by fusing genes together, so that they become attached to different *cis*-acting regulatory sites. These fusions (which may or may not involve fusions of coding sequence) produce genes that can escape normal regulation, in a dominant way, as do the Lac structural genes when fused to the Ara promoter (see Chapter 11), escaping regulation by the Lac repressor in a dominant way. This mechanism is one of the most frequently encountered changes that drive the inappropriate proliferation of cancer cells.

Another way in which genomic rearrangements produce dominant mutations is by gene amplification. In human cells, like those of the majority of eukaryotes, the expression of most genes will increase when the copy number of the coding sequence is increased, producing a dominant change in levels of expression. Again, in the setting of cancer and evolutionary change, gene duplications and further amplifications have been found to be prominent players in these processes. Transposons also play a major role in this kind of change, and their ability to facilitate duplications might explain why they have been retained in the genomes of most, if not quite all, species.

## Complex Disease Phenotypes

Genome-wide association studies (GWASs; see Chapter 14) are currently the most popular way to investigate the genetic determinants of common and complex inherited disease phenotypes. Although these studies have found statistical evidence for the enrichment of particular loci in affected individuals relative to controls, most of the gene loci implicated account for very little of the variance in phenotype. Much of the confidence in the approach derives from a few notable successes, such as the discovery, by GWAS, of the unexpected connection between age-related macular degeneration and complement factor H. Here, one of the SNPs used in the GWAS turned out to be causative with a very large effect on phenotype. The SNP is a simple missense variant, such that the risk allele encodes histidine instead of tyrosine at codon 402. Homozygotes for the risk allele are about seven times more likely to develop age-related macular degeneration and heterozygotes are about three times more likely to develop it. Biochemical studies found that the two variants bind differently to an important ligand in complement function. This variant was found in an initial study of 96 cases and 50 controls. This was truly a favorable case.

However, most GWAS results are not nearly so clear-cut. As indicated above, one finds many statistically significant associations in these studies, but with very small relative risks associated with each. Molecular biology is rarely directly helpful in such a situation, for two reasons. First, the SNPs themselves are very likely not to be the causal variants; they are just linked to them. Many of the commonly used mapping SNPs are not in protein-encoding sequences. Instead, they fall in introns or lie completely outside of transcribed regions. Second, the statistical nature of the method means that very important players in a disease may be missed simply because they are not in a haplotype with suitable, common, informative SNPs.

Nevertheless, it is possible to learn a great deal from GWASs because even with all these reservations, the genes that lie near the significant SNPs sometimes fit with a disease phenotype if one knows about their biology: The example of rheumatoid arthritis I mentioned in Chapter 14 implicated a number of genes in this condition that have a plausible biological connection to the disease.

The major drawback to GWASs today is the very large populations that are required to find signals that are statistically significant. Typical GWAS studies involve many thousands of individuals. The limitation in practice is not the ability to make the SNP determinations. Instead, it is finding enough people with the genotypes of interest for anything but the very most common diseases. A major issue that drives the numbers is a statistical issue: correcting for multiple hypothesis testing. Consider any experiment for which the expected outcome is going to be tested for significance at the conventional 5% level, which means that the positive outcome had only a 5% chance of occurring by chance. For instance, flipping a fair coin 10 times will produce 9 heads with a probability of about 0.01, safely below the 0.05 conventional level. If, however, I do this experiment hundreds of times, the probability that one of these trials will come out with nine heads is very much higher (about 0.34). In asking the question "Which SNPs are enriched in people with a phenotype?," we are interrogating each of thousands of loci, meaning that we need a very small $P$ value to be confident that this did not happen by chance. That is why the $P$ values shown in Figure 14.3 are not 0.05 but rather $<10^{-7}$.

**Gene Set Analysis:** This approach is one way around the problem of multiple hypothesis testing. In this method, genes in a biochemical, signaling, or any other biological pathway or network that are likely to be relevant to a disease are specified in advance of a GWAS study, and the study's results are assessed for significance for this ensemble, and not for each SNP separately. This procedure adds power to the analysis for statistical reasons (fewer hypotheses are being tested against the data), and a successful outcome makes it possible to elucidate the sequences for a relatively small number of loci, in the search for variants (such as nonsynonymous changes in coding sequences) that could cause changes in function on the basis of their molecular biology.

**Exome Sequencing:** Exome sequencing is a technique in which only those sequences (~1% of the human genome) that encode proteins are isolated and sequenced. This technique differs from all the others in that the search is limited to enrichment of mutations that have the potential to be causal: nonsynonymous missense changes; nonsense, frameshift, and indel (insertion/deletion) variants; etc. Here again, the number of hypotheses is much smaller than for GWASs. As sequencing becomes easier and cheaper, the use of this method will no doubt become more popular.

Finally, I need to introduce the idea that some human mutations are not inherited, but rather arise presumably in the process of spermatogenesis or oogenesis in the parents. A well-known precedent for this idea is Down syndrome, a very prevalent condition that arises because mothers, as they age, tend to produce eggs with an extra copy of chromosome 21, which, when fertilized, produce viable children with the well-characterized symptoms of Down syndrome. Most other aneuploidies do not produce viable children. In the case of Down syndrome, research findings indicate that its characteristic features are a result of some of the overproduced gene products on chromosome 21 having regulatory effects on development.

In Chapter 7, I emphasized that copy number variation is a relatively frequent event compared to point mutation. Many, if not all, CNVs involve more than a single gene. Thus, methods for measuring copy number are becoming important in understanding the basis of common human diseases.

*De Novo Mutations:* De novo mutations are mutations that appear in a child but that are not present in either parent as assessed usually by sequence or copy number measurements. Within the last decade, it has become clear that many nominally heritable complex human conditions are caused, at least in part, by such newly arising mutations. The analysis of such mutations requires, of course, the study of the DNA of mother, father, and child (so-called "trios").

A major recent success, the discovery of a number of rare mutations implicated in the etiology of schizophrenia, combined exome sequencing and gene set analysis. There were two major groups involved: One of them made a prior list of 2500 genes judged to be plausibly involved in the phenotype, and they sequenced the exomes of about 2500 affected individuals and an equal number of controls. The other group sequenced the exomes of more than 600 trios in which the offspring had the disease phenotype. In affected individuals, they found an excess of mutations in genes that plausibly affect synaptic networks, ion channels, and signaling proteins in the brain. Both studies profited from the prior discovery that copy number variation in the human genome is common, and that copy CNVs are enriched in cases over controls, not only for schizophrenia but also for autism spectrum phenotypes.

It should be clear that we are still in the early stages of understanding the basis of complex human disease phenotypes. Each advance makes me ever more conscious of how little we know about our genome and about how much more there is to learn about how it works.

## INTRODUCTORY BIOGRAPHIES

Nancy S. Wexler (b. 1945) is an American geneticist. Although she earned her PhD in psychology, she became interested in the genetics of Huntington's disease because her mother, her uncles, and her grandfather all suffered from the disease. She organized studies in Venezuela, where there are extensive families segregating the disease, and collected the pedigrees and DNA samples that were used (with James Gusella) to map the disease gene to the short arm of chromosome 4 by linkage to RFLPs. This ultimately led to the isolation of the gene and to the identification of the underlying CAG-repeat mutations. Having been at risk of the disease herself, Wexler led the efforts of the Human Genome Project to raise awareness of the importance of the ethical, legal, and social issues associated with human gene mapping.

**James F. Gusella (b. 1953)** is a Canadian-born American molecular biologist and human geneticist who isolated some of the first RFLP probes and used them (with Nancy Wexler) to map the gene for Huntington's disease. He participated in the cloning of the gene, which in turn led to the recognition that the mutations that cause Huntington's disease are the extensive tandem amplifications of CAG codons.

**Joseph L. Goldstein (b. 1940)** is an American human geneticist and biochemist who, with Michael S. Brown, elucidated the regulation of cholesterol metabolism and its connection with lipid metabolism more generally and heart disease. Goldstein's collaboration with Brown is remarkable for its productivity and duration.

**Michael S. Brown (b. 1941)** is an American geneticist and cell biologist who, with Joseph L. Goldstein, elucidated the regulation of cholesterol metabolism and its connection with lipid metabolism more generally and heart disease. Together, they transformed the way in which the biomedical world thinks about lipid metabolism and disease, a transformation that led to the development of statins, arguably the most successful drugs of the last half century.

**Alfred G. Knudson (b. 1922)** is an American geneticist and cancer biologist. His idea—that a cancer-causing gene that is inherited as a dominant phenotype from generation to generation could nevertheless act recessively in causing cells to become malignant through loss of the other allele by nondisjunction, mitotic recombination, or mutation—transformed thinking about cancer genes and introduced the idea of "tumor-suppressor" genes.

# CHAPTER 16

# What Is Next in Genetics and Genomics?

I have often been asked whether there is a difference between genetics and genomics. My answer was always the same: Genetics is, for me, a basic science, the principles of which are fundamental and invariant to technology. Genomics differs from genetics only in scale and technology. Whatever is learned with the fabulous technologies of the "-omics" era—high-throughput sequencing, high-throughput phenotyping, genome-wide gene expression ("transcriptomics"), comprehensive studies of metabolites ("metabolomics") and lipids ("lipidomics"), etc.—is data and factual information on a massive scale that, nevertheless, has to be interpreted by applying the basic principles of genetics.

A trivial personal example is illustrative: When I was an undergraduate student in 1962, spending my first summer in a biology laboratory, I was asked, as a five-finger exercise, to see whether I could assay an enzyme that degrades tryptophan (tryptophanase), which is inducible by tryptophan in the medium. I established a rudimentary assay and verified in *Escherichia coli* that indeed the activity was present only after induction. To satisfy myself that this induction was specific, I assayed tryptophanase with my homespun assay, together with β-galactosidase, under several circumstances: before and after the addition of tryptophan, and before and after the addition of TMG, an inducer of the *lac* operon (the magisterial papers of Jacob and Monod had just been published). I found that tryptophanase activity was present only after induction with tryptophan, and that β-galactosidase was present only after induction with TMG. The conclusion was that each of these inductions was "specific." I understood, at the time, that this specificity was provisional, because I had checked only one other degradative enzyme out of a potentially very large number (at that time, nobody knew how many).

Today, one would use a genome-wide expression technology (such as an expression microarray) and test not just one other inducible system but every gene in the genome. Indeed, in 2000 I participated in just such an experiment using a DNA microarray that interrogated every cistron in *E. coli*, and the results indeed showed that the two structural genes in the tryptophanase operon are the only ones induced by the addition of tryptophan. This experience, in my mind, perfectly illustrates the difference between genetics and genomics. The basic ideas behind our understanding of how degradative enzymes are induced, and how they are induced specifically, belong to the principles of genetics. What the genomics provided is comprehensiveness.

Before the "-omics" era, genes and proteins were studied one by one. Today, all the genes and proteins in an organism can be studied simultaneously, at least in principle. This is true today for gene expression studies. A comprehensive, double-mutant analysis has already been done in one model organism, and the technology for doing this in animal cell culture is just arriving on the scene. These developments augur well for our understanding much more about how biological systems work; for many processes already, the pathway metaphor is being displaced by a new one: the "network." I think it likely that as we explore biology at the "system" level, thinking about the basis of regulation, functional interactions, and feedback relationships as a network of interacting genes and gene products rather than as an ensemble of individual pathways, new principles of genetics will emerge.

In this era of "big data," the natural tendency to confuse facts with principles is greatly exacerbated. We are inundated with data. Anyone who has tried to look at the genomic sequence of an entire organism (even something as small as phage T4) knows that one can't really think about such a large volume of information. Instead, we parse the data and try to infer its structure using statistics and machine learning. But the best way to do this is to organize it around the principles of genetics, virtually none of which depends on huge volumes of data. The ideas of specificity and induction were *inferred* from a few instances (Lac, Ara, tryptophanase); all that the big data show us is that there are very many instances to be found of these processes, with the usual variations one expects from an evolved, as opposed to a designed, system.

So what of the future of genetics and genomics? Clearly, one use for genomic analysis is to find instances, as I suggest above, in which the principles of genetics, as we know them thus far, appear to be inadequate. This will then require the establishment of new, additional principles. An illustrative possibility for new principles might emerge from studies of *cis*-active sites in DNA that are very distant from the sequences they appear to regulate. How this works is as yet unclear. Another likely possibility is the phenomena of heritable regulatory states that are not due to mutations: so-called "epigenetics." Although large quantities of epigenetic data have been collected and more are pouring in, the new principles of epigenetics, like the principles of system-level networks, have yet to emerge.

I wrote this book, as I indicated in the Preface, in part because it is inevitable that each of us soon will be faced with trying to understand our own sequence. The only effective way to do this is to be armed with the principles of genetics and the language that has arisen to describe matters of principle.

## Applications of Genetics and Genomics

One of the obvious advances made possible by genome sequencing is the possibility of finding causative mutations, using these as diagnostics, and then applying our understanding of the underlying biology to devise interventions.

The case of cancer is illuminating. Clearly, the continuing search for oncogenic or tumor-suppressing mutations has led to some great successes, yet much of the promise in this field has yet to be fulfilled. This is partly because the necessary demonstration that variants implicated in human studies are actually causative is not yet a high-throughput process; it is very slow and painstaking work. Another problem is that tumors evolve, and not every mutation that causes a cancer early in its history remains essential to it later. A third, and possibly the most challenging, issue is that many causative mutations in tumors simply inappropriately express a function that is required in normal tissues: Antagonizing this function becomes a delicate business. In this situation, finding a way to inhibit the cancer without producing deleterious effects in normal cells has rarely been successful.

The notorious example here is the *KRAS* oncogene (see Chapter 15): It was one of the very first discovered, and mutations in this gene are found in most of the most common tumors. Yet despite efforts by virtually every pharmaceutical organization and a legion of academic researchers, a suitable therapy based on inhibiting KRAS has yet to be found.

Another application for genetics and genomics is in the search for drug targets. In many ways, the ideal situation is to have a drug that addresses a target that is known, from analysis of human or mouse mutants, to be dispensable with minimal residual phenotype. One example of this is immunoglobulin E (IgE) deficiency, which has a prevalence of ~3%–10%, depending on the human population studied. The biosynthesis of IgE is complicated, and many genes could result in IgE deficiency. People with this deficiency have minimal phenotypes, a fact that gave great comfort to the drug developers who produced omalizumab, a monoclonal antibody that specifically binds to free human IgE and a widely used treatment for severe allergic asthma.

Two other examples deserve mention, not because they have already yielded drugs but because they have generated considerable expectation and competition among pharmaceutical companies as potential drug targets. One of these is the

*CCR5* gene, a particular allele of which (*CCR5-Δ32*) causes a 32-amino-acid deletion in the encoded protein, a chemokine receptor. This deletion allele has a frequency of about 0.1 in Caucasian populations in Europe and North America. Lymphocytes from individuals homozygous for this mutation are resistant to infection by the HIV-1 virus; CCR5 appears to be the receptor for HIV infection. The hope is to find drugs that will block this receptor and produce essentially no side effects.

The second example is *PCSK9*, which encodes a protein involved in cholesterol homeostasis. Dominant alleles of this gene cause hypercholesterolemia and heart disease, whereas individuals homozygous for null alleles (a nonsense mutation) have abnormally low cholesterol but seem otherwise completely without phenotypic defect. The hope is that inhibitors of the PCSK9 protein will lower cholesterol and reduce heart disease without producing the limited side effects of statins, already among the most successful drugs on the market.

Finally, there is the expectation that in the future we will understand enough about our genomic DNA sequence to use it to guide our medical treatment. There has been considerable ballyhoo about this already, but it is quite clear that we understand very little about our sequence and how it works thus far. Nevertheless, over the long term, there is hope that as we learn more about the individual genes, gene interactions, and phenotypic consequences of variations in those genes and their interactions, we may indeed be able to usher in an era of personalized medicine.

# Index

## A

ABO locus, 19
Actin, 121–122, 127, 134
Activation, 67
Activator, 157
    AraC, 148–149, 151–152
    catabolite activator protein (CAP), 150
Adams, Alison E.M., 122, 128
Adaptors, 92, 104
Adenosine triphosphate (ATP), 47
AFLPs (amplified fragment length
        polymorphisms), 12
Age-related macular degeneration, 203
Agglutination of red blood cells, 19
Alkaptonuria, 185–186
Allele frequencies, 183
Alleles, 12–14
    inferring human gene function from
        disease alleles, 195–206
    mutant, 13, 20–22, 24–27
    null, 14, 26, 28–29
    pseudoalleles, 28
    wild-type, 13–14
Allelism, 12–13, 38
Allelomorph, 12
Allosteric activation, 64
Allosteric inhibition, 63
Allosteric interaction, mutations affecting, 200–
    201
Allosteric modulation, 64
Amber mutations, 110
Ameliorating interaction, 134
Amino acids
    genetic code, 87, 91–92, 92t
    in metabolic pathways, 61
    number in typical proteins, 94
Amorph, 27
Amplified fragment length polymorphisms
    (AFLPs), 12

Andrews, Brenda J., 135–136
Annotation, genome, 167
Anticodon, 92, 115
Anti-correlation coefficient, 38
Antimorph, 27
Antitermination, 149
Arabinose (*ara*) operon in *E. coli*, 148–150,
    151–152
AraC, 148–149, 151–152
Arginine biosynthesis, 124–125, 124f
Autozygosity, 186
Auxotrophs, 49–50, 53, 106–112, 131

## B

Backcross, 18
Back mutation, 105
Bacteria, coupling of transcription
    and translation, 139
Bacteriophage λ, 118–119
Bacteriophage T4
    genome, 71–73
    infection process, 71
    morphogenesis and assembly pathways, 52–
        55, 54f
    protein biosynthesis regulation, 138
    *rII* system, 72
        fine-structure mapping, 78–80
        mutations, 71, 74, 78–81, 95, 107,
            111–112, 114
        polycistronic mRNA, 144
        r phenotypes, 75
Balanced reciprocal translocations, 97
Base-pairing, 88, 92–93
Bateson, William, 8, 12, 15, 57, 133, 184–185
Benzer, Seymour, 6, 12, 17
    biography, 84
    *cis–trans* test, 81–83, 82f
    function gene definition, 81–83, 84
    phage studies by, 71–74, 78–81, 83, 95, 111

Bernard, Claude, 69
β-galactosidase, 141, 143–146, 159, 207
β-galactoside permease, 141, 143–145
β-galactoside transacetylase, 141, 143–145
β-globin genes/proteins, 20–22, 172–173
Biochemical reactions, 47. *See also* Metabolic
    pathways
Biotechnology, 157–160
Blood types, human, 19
Boone, Charlie M., 135–136
*BRCA1* alleles, 23–24
Breast cancer, 23–24
Brenner, Sydney, 114–115
Bridges, Calvin, 97, 104
Brown, Michael, 198, 206
Bypass suppressors, 124–125, 124f


**C**

*Caenorhabditis elegans*, sex determination in,
    127
CAG repeats, 199, 205–206
cAMP (cyclic AMP), 150
Cancer
    amplification of genomic sequences, 101
    applications of genetics and
        genomics, 209
    causative mutation, 10
    loss of heterozygosity and, 201–202
    oncogenes, 175, 200
    point mutations and, 175
    *TP53* gene mutation, 14
    translocations and, 98, 162
    tumor cell selection assays, 161–163
    tumor-suppressor genes, 102, 162, 175, 196,
        200, 201–202, 206
    virus integration into genome, 100
Cannon, Walter, 68–69
Catabolite activator protein (CAP), 150, 155
Catalysis, by enzymes, 47, 48, 49f
*CCR5* gene, 210
Cell cycle, 123
Centimorgan, 37
Chain-termination mutations, 96
Chaperone protein, 118
Chemokine receptor, 210
Chemostat, 62
*Chlamydomonas reinhardtii*, flagella of,
    125–126
Cholesterol homeostasis, 210
Chorismic acid, 63, 63f
Chromosomal rearrangements, 161–162, 171

Chromosomes
    gain and loss of entire, 162
    recombination, 100–103, 102f
    translocations, 97–98, 98f, 161–162
*cis*-acting elements, regulation and, 156–157,
    162
*cis*-acting regulatory sites, 177, 202
*cis*-dominance, 114, 145–146, 157
*cis–trans* test, 17, 114
    Benzer and, 81–83, 82f
    *cis*-dominance, 145–146
    Lewis and Pontecorvo, 81, 84
Cistron, 91
    *cis–trans* test and, 17, 82–83
    defined, 82–84
    polycistronic messenger RNA (mRNA), 142
Coding strand, 87, 92–93, 139
CODIS, 11–12
Codominance, 19, 21, 36
Codons, 87, 91–92, 92f
Coinfections, phage, 75–76, 78
Cold-sensitive (Cs) mutations, 119–121
Combinatorial regulation, 150–151
Combined DNA Index System (CODIS),
    11–12
Complementation, 13, 27–29
Complementation group, 28
Complementation tests. *See also cis–trans* test
    *araC* cistron, 151
    in phage crosses, 75
    suppressor mutations and, 120, 125
    T4 conditional-lethal mutants, 53
Complement factor H, 203
Complex phenotypes, 131–136
    disease phenotypes, 203–205
    epistasis, 133
    gene interactions, 134
    genetic heterogeneity, 131–132
    genome-scale genetic interactions in yeast,
        134–136, 135f
    polygenic inheritance, 132
    quantitative traits, 132
    synthetic phenotypes, 132–133
Compound heterozygote, 27–29
Conditional-lethal mutations, 52–53, 74–75, 77,
    107, 119–120, 122–123
Consanguinity, 184
Conserved DNA
    conservation of functional sequences,
        167–169
    phenotypic effects of mutations in,
        174–179

Constitutive, 141
Constitutive mutants, 141
  Lac system, 141, 143–145
Copy number variants (CNVs), 98–99, 161–162,
    171, 202, 205
Corepressors, 148, 155
Correns, Carl Erich, 15
Crick, Francis, 92, 100, 104, 114
Cross-feeding, 50–51, 51f, 55, 56
Crossing over, 31, 34, 34f, 36, 101, 102f,
    187–189
Cyclic AMP (cAMP), 150
Cyclins, 123
Cystic fibrosis, 44, 183
Cytological analysis, 97
Cytoskeleton, 121–122, 125–126


**D**

Darwin, Charles, 10, 15
Databases
  genome sequences, 167
  human mutations, 14, 175
Degeneracy, genetic code, 91–92, 177
Delbrück, Max, 84
Deletions (deletion mutations), 13–14, 175
  definition, 80
  gene knockout, 196
  in laboratory selection assays, 162
  *lac* operon genes, 146–147, 147f
  nonreverting T4 *rII* mutants, 79–80
  null alleles, 14, 26
De novo mutations, 205
de Vries, Hugo, 9, 15
Dihybrid cross, 32, 33f
Diploid, 5–6, 8, 72
Direct repetition, 101
Disease, inferring human gene function from
    alleles, 195–206
  complex disease phenotypes,
      203–205
  simple Mendelian disease phenotypes,
      195–202
DNA
  replication, 88, 157
  structure, 88, 89f
  transcription, 89–91, 90f
*dnaB*, 118–119
DNA polymerase, 153
DNA polymorphisms, 10, 11, 38–42, 39f–41f, 44,
    187, 190. *See also* Polymorphisms
  as genetic markers, 13

haplotype, 42–43
  in *TP53* gene, 14
  linkage mapping, 11, 196
  linkage phase, 42, 43f
  segregation of polymorphic loci
    single locus, 39–40, 39f
    three loci, 41, 41f
    two loci, 40, 40f
DNA recombination. *See* Recombination
DNA repair, 101, 103
DNA sequence variant, 10, 11
DNA variants, 9–14
  alleles, 12–14
  mutations, 9–10
  polymorphisms, 10–12
Dobzhansky, Theodosius, 1, 14, 132
Domain architecture of proteins and their genes,
    163–166, 164f
Dominance, 17–22, 24–27
  biological interpretation of, 24–27
  *cis*-dominance, 114, 145–146, 157
  codominance, 19, 21
  definition, 18
  gain-of-function, 25
  genomic rearrangements and, 202
  haploinsufficiency, 26
  implicit experiment and determination of,
      17–20
  incomplete (partial) dominance, 19
  loss of heterozygosity and, 201–202
  mutant allele relationship to wild-type allele,
      20–22, 24–27
  nonsense suppressors, 112
  overdominance, 21–22
  in phage crosses, 75
Dominant disease phenotypes, 199
Dominant-negative mutation, 26, 27
Dosage suppression, 122–123
Double crossovers, 37
Double helix, DNA, 88, 89f
Double mutant analysis
  in metabolic pathways, 50–52, 51f, 57
  in regulatory and signal transduction
      pathways, 55–56, 56f, 57
  as test of epistasis, 57
Downstream, 48, 139
Down syndrome, 204
*Drosophila melanogaster*, genome of, 169
Dulbecco, Renato, 73
Duplication
  gene, 27, 171–172, 202
  whole-genome, 173

**E**

*E. coli*
 as prototroph, 48
 genome size, 169
 human protein production in, 157–159
 induction-repression mechanisms in, 67
Edgar, Robert S., 52–53, 57
Effect size, 191
Eisen, Michael, 170, 179
Elongation, 154
End-product inhibition, 62–65, 65f, 66, 67
Energy, 47
Englesberg, Ellis, 148, 152
Enhancers, 151, 157
Enzymes, 47–51
 metabolic regulation at level of enzyme
  activity, 62–65
 metabolic regulation at level of enzyme
  synthesis, 65–67
Epidermal growth factor domain, t-PA, 164–165,
  164f
Epigenetics, 208
Epistasis, 57, 133
Eugenic laws, 183
Evolution, 1, 4
 as compromise, 163
 consequences of genome architecture,
  160–163
 conservation of functional sequences,
  167–169
 domain architecture of proteins and their
  genes, 163–166, 164f
 duplication and divergence of genes, 171–172
 neutral theory of molecular evolution,
  170–171
 phenotypic effects of mutations in conserved
  DNA, 174–179
 of replication, transcription and translation
  machinery, 153
 tree of life, 168f
Exacerbating interaction, 134
Exome sequencing, 204
Exons, 91, 165, 176
Experimental organisms, 3–4, 14, 182
Expression microarray, 207
Expressivity, 23–24
Extragenic suppressor, 106


**F**

Familial hypercholesterolemia, 198
Fanconi anemia, 131–132

FBI Combined DNA Index System (CODIS),
  11–12
Feedback inhibition, 61, 66
 allosteric interactions, 63–64
 example, 64–65, 65f
 in tryptophan biosynthesis, 62–63
 speed and reversibility of, 67
 usage of term, 63
Feldman, Marc, 170, 179
Fibrin, 163–165
Fibronectin finger domain, t-PA,
  164–165, 164f
Fimbrin, 122, 134
Fine-structure mapping, bacteriophage T4,
  78–80
Fisher, Ronald, 43, 46, 133
Fitch, Walter, 172, 179
5' untranslated region (5' UTR), 176, 177
Flagella, 125–126
FlyBase, 167
Fly room at Columbia University, 32, 46
Fragile X syndrome, 199
Frameshift mutations, 96–97, 112–115, 175
Friedreich's ataxia, 199
Function, 5, 6
Functional gene
 complementation and definition of, 27–29
 definition, 81–83, 84
 DNA polymorphisms and, 13
 evolutionary conservation of sequences,
  170–174
 locus reconciled with, 80–81
Functional sequences, evolutionary
  conservation of, 167–179
Functional suppression, 117–128
 bacteriophage λ, 118–119
 dosage suppression, 122–123
Functional suppressors, 109
 bypass, 124–125, 124f
 mutual interaction, 121–122
 protein interaction, 117–119
 recessive, 125–127
 with novel phenotypes, 119–121


**G**

Gain-of-function mutation, 25, 200
Gametes, 8
Gametogenesis, 33
Garrod, Archibald, 184–185
Gene amplification, 202
Genecards, 14

Gene conversion, 103
Gene duplication, 27, 171–172, 202
Gene expression, transcriptional regulation of, 137–152
Gene interactions, 134
Gene knock-in, 196
Gene knockout, 196
Genes, 12
  domain architecture of proteins and their genes, 163–166, 164f
  duplication and divergence of genes, 171–172
  functional definition, 6
  inferring human gene function from disease alleles, 195–206
  meaning of term, 6
  modular architecture of, 153–166
  number of, 169, 195
Gene set analysis, 204
Gene therapy, 100
Genetic analysis, 1–6, 8
  actin cytoskeleton, 121–122
  lactose utilization in *E. coli*, 140–141
  suppression and, 106–109
Genetic code, 87, 91–92, 92t
Genetic drift, 182
Genetic heterogeneity, 131–132
Genetic markers, DNA polymorphisms as, 13
Genetics
  applications of, 209–210
  future of, 207–210
  genomics distinguished from, 207
Genetic testing, DNA source for, 38–39
Genetic variation, 4–5
Genome(s)
  evolutionary consequences of genome architecture, 160–163
  modular architecture of, 153–166
  non-protein-encoding sequences, 169
  sequence databases, 167
  size of, 169
  whole-genome duplication, 173
Genome annotation, 167
Genome-wide association study (GWAS), 191–194, 192f, 193f, 203
Genomic rearrangements, 202
Genomics
  applications of, 209–210
  future of, 207–210
  genetics distinguished from, 207
Genomic sequence diversity, recombination and, 100–103, 102f

Genotype
  change (*see* Mutation)
  definition, 7
  penetrance and expressivity, 22–24
  phenotype connection, 22–24
Genotype frequencies, 183–184
Genotyping, 12
Georgopoulos, Costa, 118, 128
GFP (green fluorescent protein), 159–160, 160f
Gilbert, Walter, 90, 104
Globin genes, 172–173, 173f
Glutamic semialdehyde, 124–125, 124f
Goldstein, Joseph, 198, 206
Green fluorescent protein (GFP), 159–160, 160f
*groE*, 118–119
GTPases, 201
Gusella, James F., 199, 205, 206
GWAS. *See* Genome-wide association study

H
Haldane, J.B.S., 37, 45
Haploid, 6, 8, 72–73, 74
Haploinsufficiency, 26
Haplotype, 42–43, 189–191, 196
Haplotype blocks, 189–191
Hardy, Godfrey Harold, 182, 194
Hardy-Weinberg law (Hardy-Weinberg equilibrium), 182–185, 194
Hawthorne, Donald, 111–112, 115
*HBB* gene, 20–22, 172
Heat-sensitive mutations, 119–121
Hemizygote, 9
Hemoglobin, 172
  dominance relationships in β-globin (*HBB*) gene, 20–22
  hemoglobin A, 21
  hemoglobin S, 21
Hemophilia A, 10
Hershey, Alfred Day, 72, 74, 84
Herskowitz, Ira, 118, 128
Heterozygote
  compound, 27–29
  dominance and recessiveness, 17–22, 25–26
    codominance, 19, 21
    haploinsufficiency, 26
    human β-globin gene, 20–22
    incomplete (partial) dominance, 19
    simple, 17–18
  dominance via loss of heterozygosity, 201–202

Heterozygote (*Continued*)
  hemizygote, 9
  usage of term, 8–9
  verification of genotype, 18
HGMD (Human Gene Mutation Database), 14, 175
HIV infection, 210
*HLA* locus, 193
Hodgkin, Jonathan, 127, 129
Homeostasis, regulation and, 66–67, 137
Homolog, 173, 173f
Homologous chromosomes, 34, 34f
Homologous (legitimate) recombination,
    100–101, 102f, 165
Homozygosity mapping, 186–187
Homozygote
  autozygosity, 186
  usage of term, 8
Host-range mutants, 74
*HTT* gene, 199
Huang, Bessie, 125–126, 128
Human β-globin (*HBB*) gene, 20–22, 172
Human blood types, 19
Human Gene Mutation Database (HGMD), 14,
    175
Human genome, number of protein-encoding
    genes, 169
Human population genetics, 181–194
Human protein production in *E. coli*, 157–159
Huntington's disease, 199, 205–206
Hybrid organisms, 8
Hypermorph, 27
Hypomorph, 27
Hypothesis testing, multiple, 203–204
Hypoxia, red blood cell sickling under, 21


**I**

Identity by descent (IBD), 190
Illegitimate recombination, 102–103
Immunoglobulin E (IgE) deficiency, 209
Implicit experiment, 3, 17–29, 108
Inactivation, X chromosome, 178
Inborn error of metabolism, 185
Inbreeding, 184–186, 185f
Incomplete (partial) dominance, 19
Indels, 97. *See also* Deletions (deletion muta-
    tions); Insertion mutations
Independent assortment, 31–33, 41–42
Inducers
  in negative-control systems, 148
  in positive-control systems, 148–149
Inducible, 141

Inducible system, 149
Induction, 66
Informational suppressor, 109
Information flow, 88–94
Informativeness, 40
Inheritance
  complex modes of, 176
  patterns, 1–2
  polygenic, 132
Inherited disease
  genetic lesions causing, 175–176
  inferring human gene function from disease
    alleles, 195–206
Inhibition, 67
Initiation, 154, 155
Insertion mutations, 99, 103, 175
Insulin, human, 157, 159
Interallelic complementation, 29
Intragenic complementation, 29
Intragenic suppressor, 106
Intron-exon boundary elements, 176
Introns, 91, 164f, 165, 176
Inversions, 100, 101
Inverted repetition, 101
In vitro complementation, 53


**J**

Jacob, François, 66, 68, 84, 139–147, 207
Jarvik, Jonathan W., 119–120, 128
Johannsen, Wilhelm, 6, 7, 15


**K**

Kimura, Motoo, 170–171, 179
Knudson, Alfred, 200–202, 206
Koshland, Daniel E., Jr., 63, 68
*KRAS* oncogene, 200–201, 209
Kringle domains, t-PA, 164–165, 164f


**L**

Laboratory selection studies, 161–163
Lac repressor, 143, 144f, 145–148, 151, 155, 202
Lac system in *E. coli* (*lac* operon), 66, 155–157
  *cis*-dominance, 145–146
  coordinate expression of structural proteins,
    142
  deletions removing *lac* operon genes,
    146–147, 147f

functional analysis of *lac* operon regulatory elements, 144–146
genetic structure of *lac* operon, 142–144, 144f
terminology, 140–141
Law of mass action, 123
Leakiness, of mutations, 79
Lesch–Nyhan syndrome, 45
Lewis, Edward B., 81, 84
Li–Fraumeni syndrome, 14
Likelihood ratio, 46
Linkage, 31–33, 35
likelihood ratio, 46
Linkage disequilibrium, 190–191, 196
Linkage mapping, 11
disease identification by, 199
in human families, 38–43
haplotype, 42–43
independent assortment, 41–42
informativeness, 40
linkage phase, 42, 43f
Mendelian segregation, 39–40, 39f
using DNA polymorphisms, 196
Linkage phase, 42, 43f, 190–191
Locus, 12, 78
defining by failure to recombine, 38
definition and usage of term, 6
functional gene reconciled with, 80–81
measuring distance between loci by recombination, 35–37
polymorphic, 13
LOD (logarithm of the odds) score, 46, 132
Logical operators, 156
Long noncoding RNAs (lncRNAs), 178
Loss-of-function phenotype/mutation, 25, 27, 49f, 56f, 105, 197–198, 200
*araC*, 148, 152
auxotrophic mutation, 49
*lac*, 144–145
temperature-sensitive mutation, 53
Loss of heterozygosity, 200, 201–202
Low-density lipoprotein receptor (*LDLR*) gene/protein, 194, 198–199
Luria, Salvador, 84, 128
Lysis, 71

**M**

Maas, Werner K., 142, 152
Macular degeneration, age-related, 203
Major histocompatibility group, 193

Malaria, 20–22
Manhattan plots, 192–193, 192f, 193f
MAPK protein, 123
Mapping. *See also* Linkage mapping
Haldane's mapping function, 37
homozygosity, 186–187
linkage mapping in human families, 38–43
measuring distance between loci by recombination, 35–37
Maximum likelihood method, 44, 46
McClintock, Barbara, 97, 99, 104
Meiosis, 32, 34–35, 34f
crossing over, 34, 34f, 187–189
homologous recombination, 101
Mendel, Gregor
biography, 15
central insight of, 5–6
dominance and recessiveness, 17–18
independent assortment, 31–33
innovations of, 1–2
word use by, 3, 5
Mendelian segregation, 39–40, 39f
Mendel's First Law, 39
Mendel's Second Law, 40–41
Messenger RNA (mRNA), 90–91
instability of, 91, 137–138
miRNA binding sites, 178
polycistronic, 142, 148
size, 176
splicing, 90–91, 139, 165
translation, 89, 90f, 91–94
Metabolic pathways
analysis, 47–52
cross-feeding assay, 50–51, 51f, 55, 56
double mutants, use of, 50–52, 51f
terminology, 47–50
order of steps in, 50–52
regulation, 59–67
example, 64–65, 65f
at level of enzyme activity, 62–65
at level of enzyme synthesis, 65–67
posttranslational protein modification, 64
schematic, 60f
Methionine, 91
Methylation, 64
Microarrays, 190, 192, 207–208
Micro-RNAs (miRNAs), 177, 178
Microtubules, 125–126
Missense mutations, 95, 175

Mitotic recombination, 102
Model organisms, 195–196. *See also* Experimental organisms
Modular architecture of genes and genomes, 153–166
  biotechnology and, 157–160
  domain architecture of proteins and their genes, 163–166, 164f
  evolutionary consequences of genome architecture, 160–163
  initiation, elongation, and specificity in macromolecular synthesis, 153–156
  separable regulatory sites and coding sequences, 156–157
Moir, Donald T., 120, 128
Molecular biology, 87
Monod, Jacques, 63, 66, 68, 139–147, 207
Morgan, Thomas Hunt, 32, 37, 45, 104
Morphogenesis
  assembly pathways and, 52–55, 54f
  phage experiments, 119–120
Mortimer, Robert K., 111–112, 115
Mouse Genome Informatics, 167
mRNA. *See* Messenger RNA
Muller, Hermann, 9, 15, 27, 85
Multicellular organisms, genomes of, 169
Multiple hypothesis testing, 203–204
Mutant alleles
  dominance relationships, 20–22, 24–27
  null alleles, 26
  sources of, 13
Mutant phenotype, 10, 14
Mutation(s), 94–100
  auxotrophic, 49–50, 53, 106–112, 131
  back, 105
  causative, 9–10
  chain-termination, 96
  chromosomal rearrangements, 161–162, 171
  *cis*-acting, 157
  complementation analysis, 27–29
  conditional-lethal, 52–53, 74–75, 77, 107, 119–120, 122–123
  constitutive mutants, 141
  copy number variants (CNVs), 98–99, 161–162, 171, 202, 205
  databases on human, 14
  definition and usage of term, 9–10
  deleterious, 171
  deletions, 13–14, 26, 79–80, 146–147, 147f, 162, 175, 196
  de novo, 205

  dominant-negative, 26, 27
  dominant to wild type, 25
  enhancer and silencer, 151
  fixation of, 170–171
  frameshift, 96–97, 112–115, 175
  gain-of-function, 25, 200
  haploinsufficiency and, 26
  human gene, 14
  indels, 97
  insertion, 99, 103, 175
  inversions, 100
  loss-of-function, 25, 27, 49, 49f, 56f, 105, 197–198
  missense, 95, 175
  molecular taxonomy of simple, 94–97
  multipoint, 94
  neutral, 170–171, 174, 176
  nonsense, 96, 175
  nonsense suppressors, 109–112
  nucleotide expansion, 199
  null, 26, 28–29, 49–50, 49f, 96–97, 99
  nutritional, 106–112
  passenger, 196
  phage mutant phenotypes, 73–74
  phenotypic effects of mutations in conserved DNA, 174–179
  plaque morphology mutants, 73–74, 76–77, 76f
  point, 94–97, 161, 174–175
  promoter, 139–140, 146
  recessive, 24–25, 27–29, 151
  revertant frequency, 79
  RNA-splicing defects, 175, 176
  signal transduction and allosteric interaction, 200–201
  silent, 96
  somatic, 14
  strength of phenotype, 79
  substitution, 95
  synonymous, 95–96
  temperature-sensitive (Ts), 74–75, 119–121, 122, 127, 134
  translocations, 97–98, 98f, 161–162
  transposons, 99–100, 103, 202
  trinucleotide repeat alleles, 199–200, 205–206
Mutation rate, 105, 106, 182
Muton, 80
Mutual interaction suppressors, 121–122
Myopia, 192, 192f
Myotonic dystrophy, 199

# N

Nasmyth, Kim A., 123, 128
Nearsightedness, 192, 192f
Negative control, 148
Neomorph, 27
Neurofibromatosis type 1, 23–24
Neurotransmitters, 61
Neutral mutation, 170–171, 174, 176
Neutral theory of molecular evolution, 170–171
*NF1* gene, 23–24
Noncoding RNAs, 177–179
Nonhomologous end joining, 102
Nonpermissive condition, 53, 75–79
Nonsense codons, 154
Nonsense mutations, 96, 175
Nonsense suppressors, 109–112
Novick, Aaron, 62–63, 68
Novick, Peter Jay, 127, 129, 134
Nucleotide expansion mutations, 199
Null alleles, 14, 26, 28–29, 198
Null mutation, 26, 28–29, 49–50, 49f,
    96–97, 99
Null phenotype, 26, 174, 196
Nutritional mutation, 106–112


# O

Ochre mutations, 110
Ohno, Susumu, 171, 179
Omalizumab, 209
OMIM (Online Mendelian Inheritance in Man),
    14
Oncogenes, 175, 200
    *KRAS*, 200–201, 209
    proto-oncogenes, 200
    recessive, 102, 162, 200
Online Mendelian Inheritance in Man (OMIM),
    14, 195
Opal mutations, 110
Open reading frame (ORF), 94
Operator, 155
    *cis*-acting mutations, 157
    *lac*, 143, 144f, 145, 147–148, 156
Operons
    arabinose (*ara*) operon in *E. coli*, 148–150,
        151–152
    definition, 142
    Lac system in *E. coli*, 140–147, 144f, 147f
    positive control by antitermination, 149
Ortholog, 172–173, 173f, 195–196, 202
Ovarian cancer, 23–24
Overdominance, 21–22


# P

Paralogs, 65, 122, 172, 173f, 174
Parental gametes, 36
Partial diploids, 144–146
Passenger mutations, 196
Pathways, 47–57. *See also* Metabolic pathways
    epistasis, 57
    morphogenesis and assembly pathways,
        52–55, 54f
    regulatory and signal transduction, 55–57,
        56f
Patterns of inheritance, 1–2
PCR (polymerase chain reaction), 12
*PCSK9* gene, 210
Penetrance, 23–24
Permissive condition, 52–53, 75–78
Phage, 71–84. *See also specific bacteriophages*
    advantages of phage system for study, 73
    cistrons, 82–84
    coinfections, 75–76, 78
    complementation and recombination
        assessments in, 75–78, 76f
    gene and locus in T4, 72–73
    morphogenesis and assembly pathways,
        52–55, 54f
    morphogenesis experiments,
        119–120
    mutant phenotypes, 73–74
    recombination frequency, 76–77
    selective crosses, 77–78
Phenotypes. *See also specific phenotypes*
    causative mutation, 9–10
    complementation analysis, 27–29
    complex, 131–136
    complex disease, 203–205
    definition, 7
    dominance and recessiveness, 17–22,
        24–27
    genotype connection, 22–24
    haploinsufficient, 26
    loss-of-function, 25, 105
    null, 26, 174, 196
    penetrance and expressivity, 22–24
    qualitative differences in, 23
    simple Mendelian disease phenotypes,
        195–202
    strength of mutant, 79
    synthetic, 132–135
    wild type, 10
Phenotypic effects of mutations in conserved
        DNA, 174–179
    *cis*-acting regulatory sites, 177

Phenotypic effects of mutations in conserved
    DNA (*Continued*)
  long noncoding RNAs (lncRNAs), 178
  micro-RNAs, 178
  noncoding RNAs, 177–179
  protein-coding sequences, 174–176
  transcribed noncoding sequences in genes
    that encode proteins, 176–177
Philadelphia chromosome, 98, 98f, 162
Phosphorylation, 64
Phylogenetic tree, 168f
PKD database, 201–202
*PKD1/PKD2* genes, 201–202
Plaque, 74
Plaque assay, 73–74, 75–77
Plaque morphology mutants, 73–74, 76–77, 76f
Plasmid, 158
Plasmin, 163–164
Plasminogen, 163–165
*Plasmodium falciparum*, 20–21
Point mutations, 94–97, 161, 174–175
Polycistronic messenger RNA (mRNA), 142, 148
Polycystic kidney disease, 201–202
Polygenic inheritance, 132
Polymerase chain reaction (PCR), 12
Polymorphisms, 10–12, 38–42, 39f–41f, 44, 187,
    190
  amplified fragment length polymorphisms
    (AFLPs), 12
  definition and usage of term, 10
  DNA polymorphisms, 10, 11, 13, 14, 38–42,
    39f–41f, 44, 187, 190
  as genetic markers, 13
  haplotype, 42–43
  linkage mapping, 11, 196
  linkage phase, 42, 43f
  restriction fragment length polymorphisms
    (RFLPs), 11, 13, 39
  segregation of polymorphic loci
    single locus, 39–40, 39f
    three loci, 41, 41f
    two loci, 40, 40f
  short tandem repeat (STR), 11–12, 13
  single-nucleotide polymorphism (SNP), 12,
    186–192, 192f, 193f, 203–204
  variable number tandem repeat (VNTR),
    11–12, 13, 39
Polypeptide, cistron and, 83–84, 91
Pontecorvo, Guido, 81, 84, 85
Population genetics, human, 181–194
Positive control, 148–149
Positive selection, 171

Posttranslational protein modification, 64
Precursor (pre-) mRNA, 176
Pringle, John, 128
Proline biosynthesis, 124–125, 124f
Promoter fusions, 159–160
Promoters, 139–140
  human protein production in *E. coli*, 157–159
  *lac*, 143, 144f, 146, 156
  mutations, 146
    *cis*-acting, 157
    in laboratory selection assays, 162
  transcription initiation and, 154, 155
Protease domain, t-PA, 164–165, 164f
Protein fusions, 160
Protein interaction suppressors, 117–119
Proteins
  domain architecture of proteins and their
    genes, 163–166, 164f
  human protein production in *E. coli*, 157–159
  mRNA translation, 89, 90f, 91–94
  number of amino acids in typical, 94
  number of genes encoding, 169
  posttranslational protein modification, 64
  stability of, 138
  transcriptional regulation of gene
    expression, 137–152
Proto-oncogenes, 200
Prototrophs, 48–49, 107
Pseudoalleles, 28
Pseudo-revertants, 105, 108, 111–112, 120
Punnett, Reginald Crandall, 183, 194
Punnett square, 33f, 183
Purifying (negative) selection, 171, 176, 177
Purine auxotrophs, 146–147
Purine biosynthesis, 149, 155
Purine repressor, 148, 155

**Q**

Quantitative analysis, 1–3
Quantitative trait loci (QTLs), 132, 191

**R**

Random assortment, 31–33
Random drift, 171
Reading frame, 93–94
Recessive, definition of, 18
Recessive disease phenotypes, 196–199
Recessive mutations, 24–25
  complementation analysis, 27–29

enhancer and silencer mutations, 151
Recessiveness, 17–22, 24–27
    biological interpretation of, 24–27
    implicit experiment and determination of,
        17–20
    loss of function, 25
    Mendel's description of, 18
    non-absolute nature of, 21
    null alleles, 26
Recessive oncogenes, 102, 162, 200
Recessive suppressors, 125–127
Reciprocal translocation, 162
Recombinant DNA technology, 157–160
Recombinant gametes, 35–36
Recombination, 31–34, 100–103, 102f
    definition and usage of term, 33, 35
    frequency of, 35–38, 76–77, 103, 187
    gene conversion, 103
    homologous (legitimate), 100–101, 102f, 165
    illegitimate, 102–103
    in meiosis, 187–189
    loci defined by failure to recombine, 38
    measuring distance between loci by, 35–37
    mitotic, 102
    in phage, 76–78, 76f
    transposition, 103
Recon, 80
Reduction division, 34. See also Meiosis
Regulation
    cis-acting elements, 156–157, 162
    combinatorial, 150–151
    definition and usage of term, 61–62
    enhancers, 151
    homeostasis and, 66–67, 137
    initiation, elongation, and specificity
        in macromolecular synthesis,
        153–156
    at level of enzyme activity, 62–65
    at level of enzyme synthesis, 65–67
    by multiple inputs, 149–151
    negative control, 148
    of metabolic pathways, 59–67
    operon (see Operons)
    positive control, 148–149
    posttranscriptional by miRNAs, 178
    separable regulatory sites and coding
        sequences, 156–157
    silencers, 151
    transcriptional regulation of gene
        expression, 137–152, 156
Regulatory pathways, 55–57, 56f
Regulon, 142, 155

Replication, DNA, 88, 157
Repressible, 141
Repressible system, 149
Repression, 66
Repressors
    cis-acting mutations, 157
    corepressors, 148, 155
    definition, 148
    Lac repressor, 143, 144f, 145–148, 151, 155,
        202
    purine repressor, 148, 155
Restriction fragment length polymorphisms
        (RFLPs), 11, 13, 39
Retinoblastoma, 201–202
Revertant frequency, of mutations, 79
Revertants. See also Suppression
    pseudo-revertants, 105, 108, 111–112, 120
    true, 105, 108, 113
RFLPs. See Restriction fragment length
        polymorphisms
Rheumatoid arthritis, 193, 193f
Ribosomal ambiguity suppressors, 115
Ribosomal RNAs, 177
Ribosome-binding site, 158
Ribosomes, 91, 139, 153
RNA
    structure, 88
    transcription, 89–91, 90f
    translation, 89, 90f, 91–94
RNA polymerase, 90f, 118–119, 138
    as multisubunit structure, 83–84
    catalytic subunits, 153
    cis-dominant phenotypes and, 157
    promiscuity of, 155
    promoter interaction, 139–140
    RNA polymerase II holoenzyme, 156
    transcription factors, 151
RNA splicing. See Splicing
RNA-splicing defects, 175, 176
Rotman, Raquel (Sussman), 72, 74, 84
rpoB, 118–119


S

Saccharomyces cerevisiae
    as prototroph, 48
    genome size, 169
    introns, 176
Saccharomyces Genome Database (SGD), 167
Saliva sample, for genetic testing,
        38–39
Schekman, Randy, 129

Schizophrenia, 205
Segregation
    Mendelian, 39–40
    of polymorphic loci
        single locus, 39–40, 39f
        three loci, 41, 41f
        two loci, 40, 40f
Selection, 174
    extreme, 183
    laboratory selection assays, 161–163
    positive, 171
    purifying (negative), 171, 176, 177
Selective crosses, 77–78
Sequence databases, 167
Sequence homology, 100
Sex chromosomes, 8–9
Sex determination in *Caenorhabditis elegans*,
        127
*SGD* (*Saccharomyces* Genome Database), 167
Short tandem repeat (STR) polymorphism, 11–
        12, 13
Sickle cell anemia, 20–22, 172, 198
Sickle cell trait, 20–21
Signal sequence, 164–165, 164f
Signal transduction
    mutations affecting, 200–201
    pathways, 55–57, 56f, 127
    recessive functional suppression and, 127
Silencers, 151
Silent mutations, 96
Single-nucleotide polymorphism (SNP), 12, 39,
        42, 186–192, 192f, 193f, 203–204
Snapdragons, flower color in, 19
SNPs, 39, 42
Somatic mutation, 14
Spinocerebellar ataxia, 199
Spliceosome, 176
Splicing, 90–91, 139, 154, 165
    defects, 175, 176
Statistical significance, 3
Statistics, 43–45
    likelihood ratio, 46
    LOD score, 46
Stop codons, 91, 92t, 94, 154, 158
    nonsense suppressors, 109–112
STR (short tandem repeat) polymorphism,
        11–12, 13
Structure, 5, 6
Sturtevant, Alfred, 32, 45, 97, 104
Substitution mutations, 95
Suppression
    by frameshift mutations, 112–115

definition, 105–106
dosage, 122–123
functional, 117–128
genetics, 105–115
genetic test for, 109
Suppressors
    bypass, 124–125, 124f
    extragenic, 106
    functional, 109
    genetic properties of, 106–112
    informational, 109
    intragenic, 106
    mutual interaction, 121–122
    nonsense, 109–112
    with novel phenotypes, 119–121
    protein interaction, 117–119
    recessive, 125–127
    ribosomal ambiguity, 115
    tRNA-based frameshift, 115
Swiss Prot Diseases and Variants (database), 14
Synergistic interaction, 134
Synonymous mutations, 95–96
Synthetic lethality, 132, 134, 135
Synthetic phenotype, 132–135
Szilard, Leo, 62–63, 68


T
T4. *See* Bacteriophage T4
Temperature-sensitive (Ts) mutations, 74–75,
        119–121, 122, 134
Template strand, 93, 138
Termination codons, 154, 158
Termination factors, 154
Test cross, 18
Thought experiment, 17
3′ untranslated region (3′ UTR), 176, 177, 178
Tissue-type plasminogen activator (t-PA),
        163–165, 164f, 173
TMG, 207
*TP53* gene mutation, 14, 196–197, 197f
Transcription, 89–91, 90f
    basics of, 138–140
Transcriptional regulation of gene expression,
        137–152
Transcription factors, 151
Transfer RNA (tRNA), 92, 177
Translation, 89, 90f, 91–94, 139, 154
Translocations, 97–98, 98f, 161–162
Transmission coefficient, 79
Transporters, 162, 172
Transposition, 103

Transposons, 99–100, 103, 202
Tree of life, evolutionary, 168f
Trinucleotide repeat alleles, 199–200, 205–206
Trios, 205
tRNA-based frameshift suppressors, 115
tRNAs, 92, 177
True breeding, 2
Tryptophan
    auxotrophs, 62, 65, 107–109, 111–112, 131
    biosynthesis regulation, 62–63,
        65–66
    chemical structure, 63f
Tryptophanase, 207–208
Tschermak von Seysenegg, Erich, 15
Tubulin, 125
Tumor cells, in laboratory selection assays, 161–
    163
Tumor-suppressor genes, 102, 162, 175, 196, 200,
    201–202, 206
Two-factor cross, 35–37


U
Umber mutations, 110
Upstream, 48, 139


V
Variable number tandem repeat (VNTR) poly-
    morphism, 11–12, 13, 39
Vector, 158
Virus integration into genome, 100

W
Watson, James, 100, 104
Weinberg, Wilhelm, 182, 194
Wexler, Nancy, 199, 205
Wild type, 10, 13–14
Wood, William B., 52–53, 57
WormBase, 167


X
X chromosome, 8–9
    fragile X syndrome, 199
    inactivation, 178


Y
Y chromosome, 8
Yeast
    actin mutations, 121–122, 127, 134
    auxotrophs, 107–112
    conditional-lethal mutations, 122–123
    dosage suppression, 123
    genome-scale genetic interactions, studying,
        134–136, 135f
    genome size, 169
    laboratory selection assays, 161–163
    MAPK protein, 123
    network of functional relationships among
        genes, 135f
    suppressor mutations and, 120–121
    synthetic lethality, 134
    transporters, 162, 172